To Study Administration By Cases

To Study Administration By Cases

ANDREW R. TOWL

Director of Case Development and Distribution
Director of the Intercollegiate Case Clearing House
Graduate School of Business Administration
Harvard University

HARVARD UNIVERSITY
GRADUATE SCHOOL OF BUSINESS ADMINISTRATION
BOSTON • 1969

Library of Congress Catalog Card No. 77-94407
SBN 87584-080-9

Printed in the United States of America

Table of Contents

List of Exhibits

Note of Appreciation

The Visiting Professors Case Method Program about which this book is written brought to the Harvard Business School 227 deans and professors from 119 schools of business administration from all parts of the United States. By their questions and reactions while working closely with our Faculty on common problems, these visitors contributed significantly to a better understanding of education for administration by all of us. I am sure they would join me and our Faculty in taking this occasion publicly to thank The Ford Foundation whose generous support for 11 years made these programs possible.

GEORGE P. BAKER
Dean

Soldiers Field
Boston, Massachusetts
July 1969

Foreword

Unproclaimed, but of a complexity and confusion paralleling those of "the management theory jungle," with proponents and opponents equally dedicated (or bigoted) and articulate (or strident), is a "management education jungle." Among the most intensely preyed-upon creatures is the case method of instruction. The attack has usually come in the accusing form of "the blind leading the blind" or "the ignorant arguing with the unknowing" or "the case method made obsolete by the business game and management science." Such nonsense has arisen from such thickness of educational verdure that data have not been able to penetrate the thickets of the marauders. Mr. Towl's book clears away much obstructing vegetation and foliage and provides for those who wish it a shaft of light on the case method.

The most central thrust of the two comprehensive studies of academic management education published a decade ago[1] was the need for increasing the intellectual rigor of management programs, focusing them on the problems of the manager rather than on descriptive training in institutions, practices, and techniques of business. As a means toward this goal, both studies urged the utilization of the case method for some significant part of the education offered.

In line with the recommendations of the reports, the deans and faculty members of many schools became interested in learning how to use the case method. Since the case method has long been used at the Harvard Business School, it seemed appropriate for

[1] *The Education of American Businessmen,* by Frank C. Pierson and Others (New York: McGraw-Hill Book Company, 1959); *Higher Education for Business,* by Robert Aaron Gordon and James Edwin Howell (New York: Columbia University Press, 1959).

this School to contribute its experience. To this end, the Ford Foundation provided funds for a repetitive series of summer programs at the Harvard Business School to provide training in the use of the case method to faculty members from other schools. Andrew Towl was selected to be the director of these programs.

As director of these annual Visiting Professors Case Method Programs, Mr. Towl has had one of the most sustained and exhaustive experiences of which I know. He has worked with people new to (and sometimes antagonistic to) the case method. Added to this rich background is Mr. Towl's own experience of writing cases and teaching by the case method, in both this country and overseas. He is now serving as Director of Case Development and Distribution and Director of the Intercollegiate Case Clearing House at the Harvard Business School. This unusually extensive and intensive experience has given him an opportunity for sympathetic and realistic understanding of the problems faced in acquiring mastery of the method as well as of the values and satisfactions of those who have come to use it consistently.

A reader should not infer from the fact that this book limits its consideration of instruction in business adminstration to the use of the case method that the case method is a panacea for all instruction in the field, or for all types of students, or indeed for all professors. The Harvard Business School, the citadel of the case method in this field, makes extensive use of lectures, collateral readings, textual notes, library research projects, field research projects, field trips, role playing, bargaining sessions, group reports, business games, and the like. In order to put together a course, a professor must know what his students bring to the course in the way of knowledge, educational experience, and business or other administrative experience. The professor needs to know what he himself brings to the course, what he may expect as a contribution from the study of the cases and from his own conduct of the class sessions. He should employ ordinarily a combination of methods, appropriately, whether before cases are discussed to provide prerequisite knowledge or afterwards to crystallize or to challenge or to develop further ideas that have arisen from the discussion. While the enormous range of vicarious experience obtainable from the case method cannot easily be duplicated by other means, other methods can bring desirable learning not readily available from the case method.

It is apparent that much of the criticism of the case method has been based on the erroneous assumption that cases were to be used exclusively — without regard to the knowledge and sophistication of the students, without regard to the particular qualities of the instructor, without regard to the nature of the specific subject matter. The fact that the case method is a tool or technique used in conjunction with other methods appropriate to the circumstances and that the combination is planned by the faculty member undercuts such criticism. Furthermore, research developments and theoretical constructions do not replace cases; they become grist for the case discussions as they are drawn on for analysis or solution of problems.

The business game differs in that it does not become such grist; it does in fact provide an alternate method. The game affords experience to students working with other students in a common organization in a situation usually requiring repetitive analyses and decisions over periods of time. In a simulated business game, each team member is forced to live with the consequences of earlier decisions and each member experiences rather than intellectualizes the interrelationship of the various functions of a business. This experiential quality arises from the intensely emotional competitive struggle for peer-group prestige resulting from the team's performance.

Business cases, in contrast, may in an equal amount of student time provide many analytical and decision-making experiences concerned with a multitude of different problems faced by different types of management teams under differing conditions and in different industries. The ideas of many students (often of widely varying backgrounds) rather than those of the few involved in each game decision may be brought out, widening the understanding of possibilities for analysis or for decision in each of the situations. The instructor may influence the direction of class attention or introduce new knowledge or a new thought or challenge a questionable assertion to improve the effectiveness of class learning. His own understanding of class thinking is high and kept current, giving him the opportunity to vary the plan for the next or future classes. Since the nature of the student experiences from these two methods is so different, the business game and the case method are clearly not substitutes for each other, but are complementary to each other, their combined use enriching the student learning.

This book about the summer program series on the case method focuses on the use of cases. Its greatest usefulness will be to those who wish to begin or to improve their own use of case materials. Others who are interested in knowing about the case method but who do not necessarily want to become case leaders will find it a valuable source of information. The form of presentation — a simple, candid statement of what was tried successfully and unsuccessfully in the summer program series, the evolution of thinking as the series progressed, the frustrations and achievements of the participants — gives the reader a remarkable "feel" for the case method, a feel that would not have been attained, I believe, if the book had been written as a "state of the art" report at the end of the series. The very character of the book conveys something of the sense of continuing development, adaptation, and evolution which is inherent in the case method.

ROBERT W. MERRY
Professor of Business Administration

Soldiers Field
Boston, Massachusetts
July 1969

Preface

Perhaps the greatest compliment to those with whom one has worked over many years is to have so assimilated their ideas that one is no longer aware of their source. Such grateful acknowledgment also reiterates the basic theme that to study administration one must get involved in multiple relationships. Nevertheless, it is appropriate to declare that as the author of this book I take full responsibility for the facts and conclusions in it, and that neither Harvard University nor other members of the Faculty are in any way committed to the views expressed.

The range of views at Harvard is amply demonstrated, beginning with ideas of the case method in the very gracious Foreword by Professor Robert W. Merry and ending with the thought-provoking views of other colleagues outlined in Appendix 4. Sandwiched in between is my report of an intercollegiate effort of 227 deans and professors from 119 schools of business administration for whom Dean Baker appropriately expresses appreciation to The Ford Foundation. The financial support of the Foundation and the personal interest of its staff made it possible for these men from other schools, along with coaches from the Harvard Business School Faculty, to work together at a truly operational level in the study of administration by cases. The Visiting Professors Case Method Program from which this book emerged came into being largely at the initiative of the late Dr. Thomas H. Carroll of The Ford Foundation. He not only arranged for the initial grant which made the Program possible, but gave it strong personal support. His successors at The Ford Foundation continued to strengthen the Program by their personal interest in its content and by making it known to potential participants. Dr. William P. Gormbley, as the Associate Director of the Program for Economic Development and Administration of The Ford Foundation in 1962, is the one who insisted that I write about the Program. Through his allocation of the residual balance of one grant, I was given the opportunity to visit alumni of the Program and an opportunity to organize the material.

It was George F. F. Lombard, Senior Associate Dean, who provided time for writing by taking on additional tasks personally as well as providing assistance. Of even more importance was his encouragement, nearly 20 years ago, that I undertake direction of the Program. Discussions with him during every stage of the entire Program as well as throughout the project of writing were of continuing support. He participated frequently in the Program and significant excerpts from some of his statements add much to the meaning of this book. He also secured the critical comments of anonymous readers thus helping to shape the final draft.

The final writing also was directly influenced by the detailed comments on earlier drafts by Dr. Melvin T. Copeland and Dr. John C. Baker, both of whom have been an inspiration to me for more than 35 years, first as teachers, then as friends and colleagues. Professor James W. Culliton also read an earlier draft and his copious notes helped to remove numerous ambiguities. Professor Fritz J. Roethlisberger repeatedly provoked new points of view and identified distinctions I was trying to make not only in drafts of the book, but in my earlier interpretation of notes. Dr. John I. Reynolds and participants in a Latin American case method program in the summer of 1968 were most helpful in clarifying several of the chapters by trying to put the ideas into operation.

At least 12 predecessor efforts contributing to this Program are summarized in the introductory chapter because no prefatory acknowledgment would adequately reveal their contributions to this book. The 227 individual participants and the 54 outside speakers, who are listed in the appendix, all made significant contributions to the Program and to the ideas in this book. Especially to be acknowledged as contributors are the 28 colleagues who served as the coaches. They were the ones who worked with the visiting participants through the whole complex of activities covered loosely by the term *case method.* Professor Thomas C. Raymond served as coach for two programs and also as director of the Program in the summer of 1961. The efforts of these coaches to make visiting professors feel like professional colleagues even while taking the student role were significantly reinforced by other members of the Harvard Faculty and especially by Dean and

Mrs. Stanley F. Teele and Dean and Mrs. George P. Baker who offered their hospitality at the Dean's House and garden.

Much of the way I have organized the book and described the activities of the Program took shape in attempting to repeat the essence of the Program in India, in Turkey, and in Mexico where the effort to make meaningful translations in different cultures and into Turkish and Spanish refined ideas and their expression. Especially in the relative isolation of these assignments my wife, Merrilie McAfee Towl, helped to recall the reality of the complex relationships with participants over all the years. She had known all of them at Harvard and had visited many of them with me in their own universities. Perhaps a token of her dedication and contribution to this book was the weight the portable typewriter for work on this manuscript took in place of exotic souvenirs and the time spent on transcribing my dictation when she might have been sightseeing.

My secretary, Mrs. Jean Burleson, typed letters, memos, and schedules for administration of succeeding Programs. Her care in maintaining records during the Programs and then resourcefully finding materials for the book provided much of the helpful detail. Her willingness to type, retype, cut, and rearrange endlessly as the mass of material slowly took shape was a creative part of the whole project.

Miss Hilma Holton, Associate Editor of the Division of Research, has not only edited the book for publication but skillfully helped in major revision of the final draft. I am indebted to her and to Professor Lawrence E. Fouraker, Director of the Division of Research, for taking on the work of this publication although it did not grow out of a research project of the School. When the printer's page proof fortunately stopped the author's efforts, Miss Josepha Perry helped to show by her skilled indexing how everything is related to everything else in the study of administration which is the end of this case.

ANDREW R. TOWL

Soldiers Field
Boston, Massachusetts
October 1969

Chapter I

The Case Method: A Means, Not an End

Administration — the process by which organizations initiate and adapt to change — is so critical to survival that any means to its improvement merits careful inquiry. Heroic efforts to find accommodation among the national and international organizations of the world headline the front pages as the financial pages record the innovations by which science and industry serve society. While scholars seek the laws by which these complex forces work, those responsible for decision at any moment are keenly aware of their need for skill as well as knowledge in charting a new direction for an organization. Those responsible for selecting and persuading men to take such positions are keenly aware how difficult it is to find or to develop men with this orientation and the necessary emotional maturity. Clearly no book is going to throw much light on such a complex tangle, but the urgency for man to find a way through it calls for any light which may have been found.

Thus we report for active administrators, potential administrators, students, and professors of administration about a program to study administration in business by cases. The Visiting Professors Case Method Program, with the support of The Ford Foundation, brought together 11 groups of 20 professors and deans from schools of business administration for 8 weeks each summer from 1955 through 1965 at the Harvard University Graduate School of Business Administration.

Looking at this Program may have novelty for active administrators and for students because the principal actors in the Program were professors. They may see, however, that in this Program the professors were committed to participate concurrently in a whole set of activities and relationships of significance to others:

to seek out and understand actual situations from the administrative point of view,
to relate such concrete situations requiring action to the abstract knowledge and concepts of the subjects they taught,
to use such situations with students in the process of learning and maturity.

This Report and the Audience for Whom It Was Written

The Program was *not organized* as a research project but as a working model of the case method, refined from year to year under the prodding and probing of a cross-section of professors from different subject and university backgrounds. Year by year the reactions of each group led to a clarification, not only of the means of communication used in the Program, but also to a clearer understanding of what "the case method" means. Therefore, this report is primarily an effort to continue the process of communication and clarification so that others may see for themselves, as much as words will permit, what these 227 participants from 119 schools found. The task of developing appropriate methods for the study of business administration is not finished. It has just begun. Practitioners as well as professors will be involved increasingly in the years ahead. Students also have a stake in learning how to study business administration, because the knowledge of today will soon be obsolete.

The purpose is to evoke the reader's involvement and to give him some of the feedback he might have gotten if he had been participating with others in the Program. This book is not an abstract theoretical model but a mixture of working conventions, rules of thumb, practical distinctions, practices, and interpretations. It invites those who would like to have more order and consistency to join in the exploration of the entire set of activities and relationships. The apparent clarification one may make by simplifying assumptions or by omitting an activity or relationship merely shifts the burden to demonstrating that the oversimplification would be relevant under the conditions of interdependence in an actual set of activities.

A major reason for the Program in the first place was that it is very difficult to translate a statement about the case method into action, and vice versa. Over these 11 years persistent attempts were made to state the ideas more clearly in the hope that, by clarification of the concepts, the use of them would be made easier. Nevertheless, it is manifest from the experiences of these several years that no amount of verbal clarification will take the place of actual use of the ideas. The 11 Programs were variations on the common theme of converting abstract ideas about the case method into a living pattern. Reading a book is not a substitute for actually studying administration by cases. The Program was structured each year upon the belief that understanding the case method depends upon completing at least one cycle of experience: in the field situation, in the writing of a case, in the use of a case in a course, and in reflection upon the gap between one's abstract ideas and the concrete situation. The insights and implications from the personal involvement required in the study of administration go beyond what can be communicated at a purely verbal level. The contribution of this book will be in stimulating further experimentation rather than in evaluation of a program now past.

The uniqueness of the Program and of this book is emphasis upon the totality of the set, the interdependence of activities and relationships in the study of administration made possible by cases. Individual readers, as did individual participants in the Program, may know far more about the facts of a particular enterprise, the logic of a particular subject, the laws of a special discipline, the nature of learning, the process of maturity, or even the case method or some aspect of a situation, than the one responsible for action in it, or the case researcher — definitely more than the writer of this book. The objective is not omnipotence, but involvement in a continuing process.

While business administrators are unavoidably immersed in the process of adapting institutions to the changing needs of mankind, the rest of modern society can ill afford to ignore it. Doctors, lawyers, farmers, clergymen, housewives, and congressmen are all involved in the system of institutions which shape our lives. Administration of change is a critical function not only in business organizations but also in hospitals, churches, government agencies, and universities. And in developing countries the adaptation of imported institutions symbolic of "development," to the needs of local culture, finds lack of the knowledge and practice of administration to be a critical block. The need for better management or administration exists whether the economic system is private, public, or mixed. In this book emphasis is upon administration in business because business and business schools were the context of the Program.

The apprenticeship method of learning business administration long ago became inadequate, and intensive efforts are now devoted by business to executive development programs. However, concerted study of the administrative process in universities is surprisingly new. Even in professional schools of business administration, most professors are men who had their own graduate study in other fields. They specialized in related but different subjects, such as economics, psychology, history, government, industrial engineering, or statistics. Much has been written about various aspects of the process or about administration in the abstract. The basic ideas are relatively simple but the implications are far-reaching, complex, and interrelated.

A significant level of discourse is needed, a level at which communication can be achieved among practicing administrators, potential administrators, and students and professors of administration and other relevant disciplines. To study business administration, one needs to get the data; to separate the data from his own thinking about the data; to relate the data and his thinking about them to his general conceptual framework in a way which is subject to criticism by his peers; and to test ideas by action in concrete administrative situations.

Out of this common need for a statement of data emerged the idea of *the case* — the facts in a concrete situation creating an issue requiring discretionary action. Such a case provides a body of data relevant to the administrator directing a company, to the scholar testing his concepts, and to the student developing orientation to administrative responsibility, knowledge, skill, and maturity.

Efforts in the Program were concentrated on the discovery of such cases in concrete situations. Further, the Program used these cases in course design and class discussion. These efforts revealed the need to strike a balance among a number of competing values creating tension, and the need to go back to other concrete situations, to search for uniformities and differences. Thus, by such successive approximations, progress was made in the field, in case writing, in course development, in classroom discussion, and in identification of career objectives.

This process has its frustrations. During the Program participants often became irritated because the coaches seemed so noncommital on any question or proposition about the case method, but were adamant about keeping on with the schedule. The response to most questions in effect was, "Let's see what difference that view makes in getting data from the field," or "in designing the course," or "in using the case in class," or "in grading students." To the participant who had had experience in administration, the advice was, "Look at the classroom situation." To the experienced teacher, the advice was, "Look at the concrete situation in the field." *Then,* "Let's consider how one develops a *course* using cases." The Program was "permissive" as to any observation but "normative" as to the schedule, urging that participants reconsider such observations in other parts of the total set of activities and relationships and share such feedback with the total group. Paradoxically, the schedule kept the participants enlarging their involvement in this interdependent set of activities and relationships which in turn forced participants into choices making them increasingly aware of their own feelings and beliefs about knowledge, skill, professional orientation, and personal growth. Questions were thus answered by providing new questions.

Before moving into the details of the Program for the study of administration by cases, brief attention to the preconditions and predecessors of the Program may help to establish a frame of reference for understanding it and for transfering ideas to other situations.[1]

Preconditions and Predecessors of the Program

The contributions of our predecessors are as critically important for the future as they were in the past. Some efforts in the last decade have been frustrated by inattention to fundamental conditions long since almost forgotten. The debt owed to Wallace B. Donham by all who are interested in the process of administration and in learning is beyond calculation. Under his leadership as Dean of the Graduate School of Business Administration at Harvard University from 1919 to 1942, the case method became almost the trademark of the Faculty. It does not detract from his genius to note that essential ingredients of educational policy had been established before he became Dean because, without recreating these conditions upon which he built, the subsequent efforts of others may well be thwarted. Credit for design of the Program might well begin with the businessmen and educators who, before 1908, more than 60 years ago, insisted that *a separate Faculty* committed to the study of business administration be established at Harvard University. Fortunately, an authoritative perspective on this early evolution has already been prepared by Dr. Melvin T. Copeland, *And Mark an Era* (Boston: Little, Brown & Company, 1958). Dr. Copeland as teacher, supervisor, and colleague is indeed the pioneer in the study of administration by cases.

From the earliest efforts to incorporate the study of business

---

[1] As the one who was privileged to be called Director of the Program in 10 of its 11 years and as the author of this book, I take responsibility for this statement, but those colleagues, fellow participants, and guests listed in the Appendix whose direct contributions were so important in the success of the Program from year to year would join me in acknowledging the roots from which the Program grew.

administration into the university at least these following fundamentals must be recognized:

Commitment to Study Administration

It is significant that the early thinking about development of the Harvard Business School grew from awareness that administration in business and government had become a fit subject for scholarly inquiry. Therefore, a separate Faculty was created for the study of administration in business.

The Discussion Method

In this new Faculty, it was soon recognized that new methods appropriate to the subject had to be found. As early as 1910, for instance, Dean Gay impressed upon Dr. Copeland that, as a new instructor, it was not enough to concentrate upon what he was preparing to say; he should plan how he as the instructor might get the students drawn into discussion of the subject.

The Emergence of Business Functions as well as Institutional and Industry Subjects

In the early stages of its evolution, this new Faculty was struggling to find, just as present faculties in schools of business administration are struggling to find ways of subdividing the work of research and instruction. Support from business seemed to come more readily at first in terms of industries. The Faculty, however, was even then searching to find a more effective common denominator. In the decade 1909–1919, Dr. Copeland observed that the Faculty had recognized the functional areas of marketing, production, finance, and accounting as a more effective basis for subdividing the curriculum than by a series of courses on different industries.

Executive Problem and Functions

In this early period the Faculty recognized that, regardless of industry or function of the enterprise, executives had on their desks problems which had considerable similarity. Executives were invited to come into class and to describe the problems they were facing. Students were asked to make a written analysis of the problem which the executive would review. He would later return to discuss with the class the respective analyses and action proposals.

Written Cases

The written case which has come to symbolize the case method actually appeared rather late in process. Note that in the first decade a new Faculty in a new Graduate School of Business Administration had discovered an interrelated set of activities, fundamental ingredients for the study of business administration. They were already aware of using many elements in the set of activities later integrated into a case method pattern. All of these elements had been drawn into the collective pool of experience within the Faculty before the challenge was made to prepare written cases. This development was described by Dr. Copeland as a request from the newly appointed dean, Wallace B. Donham, for him to shift to publishing a collection of cases instead of completing a book in marketing that was almost ready for the press. The new dean, having been trained as a lawyer in the case method, was convinced that a similar development of written cases would make the new Faculty of business administration more effective. Dr. Copeland was willing to try. The result in 1921 was the first book of written cases.

Although the elaboration of the case method continues, its elements go back to the original commitment to the study of business administration at firsthand and to the economics of getting large numbers of students to share in this type of learning. Some further steps are useful to note.

Organization for Case Research

The Bureau of Business Research, which in 1920–1925 was under the direction of Dr. Melvin T. Copeland, was the center for the next efforts at case development. Part of the effort was directed to the accumulation of cases which were then expected to become precedents, in the legal tradition. These cases were, in fact, bound in the same materials and format as the typical case books in a law library.

The organization of the Bureau in the early 1920s with a "straw boss" and an editor for each team of three case researchers established high literary standards and effective conventions of case writing. Out of this early organization came many of the statements regarding style and literary expression used during the Program and currently in use at the Harvard Business School. Also, the current system of files, numbering, and other procedures are deeply etched with the decisions made in this early period.

The excellence of the standards and procedures of a highly organized central bureau, however, conflicted with the academic tradition that each professor develop his own style and materials. Even today, when a new professor takes over a course, he tends to substitute his own cases as quickly as possible in place of his predecessor's even though they had been considered to be the last word in excellence. In 1925 case development was removed from the Bureau of Business Research and became the responsibility of individual Faculty members, who were charged with preparation of materials for their own courses.

The Case Research Team: Professor and Research Assistants

The practice, begun in the early 1920s, of hiring recent graduates of the MBA program to spend a year or two in case writing for different Faculty members has continued. The young research assistants have contributed not only additional time and energy for field work but, more importantly, have allowed professors to criticize the cases which came out of this process in a way which few Faculty members can do with their own writing. (This fact is of particular importance in other institutions when considering transfer to the case method.) Periodically the stresses and strains of working with these younger men came to general Faculty attention. Even though these recent graduates had studied hundreds of cases as students and written "distinction" reports about cases, they almost universally required from three to six months of painful trial and error before they were able to write acceptable cases from field situations. Excellence in case analysis did not automatically produce excellence in case research and writing. Members of the Faculty varied considerably in their willingness and ability to help research assistants accelerate this process of learning to write cases.

The human problem of the young men struggling to satisfy ambiguous performance criteria, and the financial problem of the School in getting results from what has always been a high-cost operation, have periodically brought administrative and Faculty attention to the problem of training case researchers. One of these efforts was an important antecedent of the Visiting Professors Case Method Program.

The WAC Task Force: Learning by Doing — with a Coach

In the summer of 1953 an experiment was started in the course which is now called "Written Analysis of Cases." Mr. James Surface, later Provost at the University of Kansas, was asked to supervise two new research assistants, and the experience of these three men turned out to be a key step in the Program. Mr. Surface himself had spent the previous year as a research assistant writing cases in the course. When the experiment was started, he agreed to keep a log of his activities during the case writing training period.

This log was important intrinsically because it was a faithful record of the activities that were involved in getting out into the field to discover and write up a case. It also identified some of the tensions in the interaction between the man in the field and the usually more senior man who was trying to utilize the observations from the field as case material in a course for which he had responsibility.

Mr. Surface's success with supervising these two new case researchers, somewhat in the old tradition of the section boss, came to the attention of a Faculty Committee on Research Policy. They encouraged the Dean to enlarge this Task Force for the training of other Research Assistants. Subsequently, task forces were organized, each with a member of the Senior Faculty assigned to work as a coach with three or four new research assistants during the summer. One purpose was to get cases for the course in Written Analysis of Cases. Another, of equal importance, was to provide supervised training for new Research Assistants as they were trying to shift from the study to the writing of cases.

Observation of this WAC Task Force in 1954 led Dr. Thomas Carroll of The Ford Foundation to propose that visitors from other schools be allowed to participate.

Mr. Donham's Visiting Firemen — Intercollegiate Cooperation

Another concurrent lineage must also be recognized. The impetus given to development of the case method by Mr. Wallace B. Donham, Dean of the Harvard Business School from 1919 to 1942, has already been noted. His determination to get across a point of view and a curriculum to implement it was not without adverse reactions at Harvard and elsewhere in the academic community. But by the time of his retirement, Mr. Donham's contribution to education was such that President Conant of Harvard University asked him to develop a course in "Human Relations," using the case method, and designed for undergraduate juniors and seniors. The fortunate aspect of this assignment from our present point of view is that it was the occasion for a number of other universities to send members of their faculties to work with Mr. Donham on the project.[2] Before this period and overlapping it to some extent, two or three professors or deans from southern schools had been coming to the Harvard Business School for an academic year under grants from the General Education Board of the Laura Spelman Rockefeller Foundation.

[2] Dr. John C. Baker, President Emeritus of Ohio University, has called my attention to the fact that when he was an assistant to Dean Donham in the 1920s, the idea of a program for university professors was already actively being promoted. In the summers of 1927 through 1931, courses were given for business executives, and professors from other universities were included so that they might learn about the case method. The trauma of the Great Depression and World War II almost erased the record of this important predecessor of the Case Method Program.

Wallace B. Donham and Elton Mayo held weekly seminars with these groups of visiting professors and some of the Harvard Business School Faculty. Those from other universities soon came to be known as "Mr. Donham's Visiting Firemen." Professor George F. F. Lombard provided the integrating administrative spirit in getting visitors involved in case writing and classroom discussion of the new undergraduate course in Human Relations. The multiple concern of this seminar group for understanding human relations, general education, and the case method when they came from backgrounds of mathematics, history, biology, literature, economics, psychology, and sociology, put the case method into the broad perspective of the whole teaching-learning process. Furthermore, the immediate post-World War II surge of students found Professors Elton Mayo and Fritz Roethlisberger and their associates turning from intensive research in industry to the problems of communicating a point of view. The seminars with Mr. Donham's visiting firemen provided an opportunity to draw many Faculty members into wide-ranging discussions. When these visiting Faculty members returned to their respective schools, the tremendous vitality of these seminars spread across the country. "Alumni" of this program wanted to come together periodically to compare notes. In the annual reunions, Mr. Donham's visiting firemen were rapidly outnumbered by colleagues who wished to join them. Exhibit I-1, prepared January 5, 1955, by George F. F. Lombard, shows the spread of this group to 38 academic institutions and 7 industry groups.

By 1955 the various related but different emphases in this expanding body were becoming somewhat more clarified. Some participants were preoccupied with the problems of general education in undergraduate colleges. Others were anxious to get on with the systematic study of organizational behavior, and another group was concerned with the case method. While the interrelationships of the learning process, the scientific study of organizational behavior, and the case method were recognized, the time had come when concentration upon different phases of these common interests could be beneficial.

Mr. Donham's visiting firemen were clearly a very significant antecedent to the intercollegiate effort at studying the case method. Specifically, the experience of Professor Lombard and the late Hilden Gibson of the University of Kansas in organizing continued cooperation among professors from different universities for exchange of cases provided the concrete format of an initial case bibliography and the idea of a clearing house.

EXHIBIT I-1

NUMBER OF PERSONS ATTENDING CONFERENCE OF MR. DONHAM'S VISITING FIREMEN

Year	*Location*	*Number*
1946–47	Harvard	20
1947–48	Harvard	24
1948–49	Colgate	34
1949–50	Ohio	120
1950–51	Kansas	70
1951–52	Harvard	108
1952–53	Colgate	75
1953–54	Ohio	107
1954–55	Kansas	98

(Total Attending 1 or more Conferences = 287)

Clarification and Publication of Concepts

The renewal of instruction after World War II with new curriculum, new courses, and new cases at the Harvard Business School, along with the spreading interest from Mr. Donham's intercollegiate seminars, brought demand for statements about the case method. Two different books were conceived and eventually brought into being.

On November 22, 1948, Harvard Business School Professors James W. Culliton, Malcolm P. McNair, and Stanley F. Teele volunteered as an informal committee to explore the possibilities of preparing a new volume on the case method to replace one edited by C. E. Fraser and published in 1931.[3] In a notice to the Faculty on November 22, 1948, they said:

It is our early thought that such a volume should be truly

[3] *The Case Method of Instruction: A Related Series of Articles,* by C. E. Fraser (New York: McGraw-Hill Book Company, 1931).

the expression of thinking on the subject at the Business School and that it should therefore be made up of chapters or sections written by many men. Different approaches, different modes of expression, seem to us desirable, if not necessary, to describe the rich fabric of the case method.

All Harvard Business School Faculty members were invited to propose topics on which chapters should be written and to indicate which of the topics they would be willing to undertake. Likewise, two prizes of $500 were offered: one for the best article entitled "The Case Method as a Research Assistant Sees It" and one for the best article entitled "The Case Method as an Instructor Sees It." This contest was initiated to make sure that younger members of the Faculty were encouraged to take part in the common effort.[4]

At about this same time, the interest spread by Mr. Donham's visiting firemen led to the request for more statements specifically on the problems of using the case method in the teaching of human relations and administration. Professor Kenneth R. Andrews undertook to edit a collection of papers[5] which were prepared in part for the annual conferences of these visiting firemen and their colleagues.

These efforts of colleagues at the Harvard Business School to put their experience with the case method into print are another obviously important contribution to the Visiting Professors Case Method Program. There is little to be said about the case method which has not already been said in these papers. They do derive from a particular phase in the development of the case method and from a particular institution. The subsequent Visiting Professors Case Method Programs had the benefit of being organized from the point of view of a cross-section of educational institutions. In the process of being repeated eleven times, these Programs also eliminated some of the localized conditions in favor of a clearer perspective of a generalized set of activities for the study of administration by cases.

---

[4] *The Case Method at the Harvard Business School,* by Malcolm P. McNair and Anita C. Hersum (New York: McGraw-Hill Book Company, 1954).

[5] *The Case Method of Teaching Human Relations and Administration,* by Kenneth R. Andrews (Cambridge: Harvard University Press, 1953).

The indebtedness of the Visiting Professors Case Method Program to these two previous books includes the personal transfer of learning that took place when Professor James W. Culliton came back to serve as one of the coaches in the first year of the Visiting Professors Case Method Program, after having gone to the University of Notre Dame and becoming Dean of their College of Commerce. Professor McNair also served as a coach in the 1956 combination of the Visiting Professors Case Method Program with a special retailing task force. Many other contributors to these books became coaches or led seminars in the Program. Finally and of prime import, Professor Teele, when he became Dean, was instrumental in inaugurating the Visiting Professors Case Method Program.

The Case Development Policy Committee: Recognition of the Case-Course Complex

Even as these books were being published, Donald K. David, then Dean of the Harvard Business School, appointed a Faculty Committee on Case Development Policy under the chairmanship of Professor Edmund P. Learned. Its initial meeting was on October 2, 1953, and the committee met almost once a week for nearly three academic years, with individual members of the Faculty or small groups, to discuss experiences in case development. Its first report on December 16, 1955 (after the summer of the first Program), summarizes the basic view of the Visiting Professors Case Method Program.

> As the Committee proceeded with its review, it found that it was unrealistic to think of cases and case writing as a distinctly separate School activity. We found that a new case became useful for an instructor only when it fitted into his conceptual framework for his course or when it challenged an instructor to reconstruct the conceptual framework for his course. In other words, the case writing process was found to be so intimately bound up with the process of course development that it seemed essential to study the development of *the entire case-course complex.*" [Author's italics.]

The deliberations of this Committee clearly contributed to the flow of ideas and interests out of which the Visiting Professors

Case Method Program developed. The careful analysis made by this Committee also was immediately helpful in identifying the conditions aiding and hindering case-course development. The willingness of the Harvard Business School Faculty to accept the suggestions of The Ford Foundation to conduct a program for visiting professors came not because the School had developed and mastered the case method of instruction but because, in the words of the Committee: "The Committee feels strongly that to date the School has by no means exhausted the potential educational possibilities of the case method."

The Ford Foundation's Professional Interest and Financial Support

Meanwhile, in the spring of 1953, Dr. Thomas H. Carroll, Vice President of The Ford Foundation, was stressing with the Foundation the importance of supporting case research. He had a long background of interest in the case method. He had been an Assistant Dean when Wallace B. Donham was Dean of the Harvard Business School. Subsequently, he had been Dean at Syracuse University and then Dean at North Carolina before going to The Ford Foundation. Also, in the summer of 1953, Dr. Carroll visited the Harvard Business School when the WAC Task Force, a predecessor to the Program run for our Faculty researchers, was having a review of some new cases which had just been completed. At that time he discussed the possibility of including visitors from other faculties in some such program.

Fortunately, the outcome of a large number of discussions between members of the Harvard Business School Faculty, administration, and representatives of The Ford Foundation led to a grant to Harvard University from The Ford Foundation on November 1, 1954, for "exploratory research and advanced training activities" including "encouragement of the 'case method' of instruction and case gathering activities throughout the business schools of the country."

To further this activity, Mr. Oscar Harkavy of The Ford Foundation visited the Harvard Business School on Wednesday, January 19, 1955, and met with a number of Faculty members at luncheon. The staff of The Ford Foundation continued to take a personal interest in the resulting intercollegiate effort. Mr. Harkavy's minutes of the meeting, which were circulated, helped to state the issues in a way which facilitated widespread discussion within the Harvard Faculty and with representatives from other schools. Dean G. L. Bach of Carnegie Institute of Technology, for instance, wrote, "However much faith one has in a case approach, a thorough exposure to, and understanding of, intelligent criticism of it seems important to effective teaching." (Dean Bach, serving as a constructive critic, kindly came to the Program several times, as will be reported later.) It became increasingly clear as the winter went on that more was involved than three or four visitors to be annexed to the WAC Task Force.

By the middle of March 1955, at the request of Associate Dean Stanley F. Teele, Professors Franklin Folts, Director of the Middle Management Program, Thomas Raymond, in charge of the WAC Program, Pearson Hunt, and Andrew Towl drafted "an experimental program" based on five major ingredients (subsequently Professor Paul Donham and Professor James Culliton of Notre Dame were added to this first "task force"):

(1) An opportunity for each participant to observe and analyze instruction by the case method.
(2) An opportunity to collect at least one case.
(3) An opportunity to teach his own case.
(4) An opportunity to develop a course of his own by the use of cases.
(5) An opportunity to participate in the grading of student reports on a case assignment.

Soon the die was cast. Letters of invitation were sent out at the end of March 1955 to members of the American Association of Collegiate Schools of Business to nominate a faculty member. In subsequent years the invitations were open to all collegiate schools of business. There never was a printed brochure, as the whole atmosphere was that of a small group of colleagues working together on common problems of studying business administration by the case method.

The existence of the new Middle Management Program with regular classes in session during the summer and a group of

Harvard Business School Faculty members actively working with the problem of course construction was a central point around which the original program was planned. However, the Middle Management Program ceased operation during the summer after the first four years of the Case Method Program.

In retrospect, not enough recognition was given at that initial stage to the strategic role of the coaches. Without them there could have been no Program. They are listed in Appendix 2, and their contributions appear repeatedly throughout the book. Their skill in "leadership from alongside" enriched and implemented the Faculty's tradition of helping visiting firemen learn by doing and discussion.

The preliminary arrangements of the Visiting Professors Case Method Program eventually were all accomplished and the lead paragraph of the press release for the morning papers of June 20, 1955, read:

> BOSTON — Deans and ranking Professors from 18 business schools representing 15 states have assembled at the Harvard Business School to begin today an experimental 8-week summer program of instruction on the use and writing of cases in teaching business administration.

In summary, by 1955 a format of organization had emerged, statements about the case method were being formulated, and an intercollegiate interest had been cultivated. To these ingredients The Ford Foundation, through Dr. Thomas H. Carroll, Vice President, and his associate, Mr. Oscar Harkavy, added professional encouragement and financial support. The result was an eight-week intensive effort for eleven summers to study administration in business by the case method.

The Program in effect became an operating model or laboratory. From 15 to 25 professors went out into business each summer to get the data of administration, firsthand, by observation and interview. They struggled to see what the facts meant, carefully keeping their opinions separated from the data which they wrote down in "cases." These and other cases were analyzed and discussed by the group, leadership rotating among the participants. As the stock of cases which each man knew enlarged, he gave attention to how he might design from these and other cases a sequence of cases and conceptual statements to meet his own course objectives. In the process, the objectives themselves took on new dimensions, for the study of administration in business by cases not only leads to reexamination of the whole administrative process but also the process of learning and maturing.

THE PROGRAM IN BROAD OUTLINE

The formal announcement of the Program was sent to the deans of 84 schools of business administration by the Dean of the Harvard Business School. Deans were personally invited each year to send faculty members to the Program.

The following excerpts from the letter of invitation to the 1965 Program show how activities were presented to prospective participants.

> The nominee does *not* need to accept the case method, and critical evaluation and analysis of different case methods will be encouraged by many seminars and outside speakers with different points of view. Nevertheless this is *not* a program of talking *about* the case method. It is a program of discovery and development by direct experiment *in* the case method, including field research in companies, writing one's own case, analyzing cases for class discussion, selecting cases for courses, and leading case class discussions. In these experiences from day to day men must make choices. They are not bystanders or observers, but direct participants in many different roles. Maximum effort is made to provide feedback from these choices by organized conferences with one's professional colleagues as well as through the informal discussions which go on continuously.
>
> . . . . .
>
> To facilitate these activities, the Harvard Faculty think of themselves as coaches, not as instructors. A coach works with a team of five or six participants to set up case leads in order that the field case research can be started very early in the program. This is necessary to allow maximum time for preparation of cases and presentation of them to the group. The coach also works with members of his team in discussions of case research, case preparation, and case discussion leadership. We try to bring

together as coaches men with quite different backgrounds, so that within the program as a whole it should be clear that there are many case methods rather than any single stereotype.

Although each day was different and over the eight-week period the mix of activities varied, participants agreed that an eleven hour day, for five days a week, was the average time needed, and that it was distributed about as follows:

(1) Case analysis and designing a course of action (about 3 hours a day)
(2) Classroom case discussion and leadership (about 2 hours a day)
(3) Field work and case writing (about 3 hours a day)
(4) Course development (about 1 hour a day)
(5) Supporting seminars (about 2 hours a day)
  (a) The Professor and His Students
  (b) The Professor and His Subject
  (c) The Professor and the Enterprise under Investigation
  (d) The Professor and His Case
  (e) The Professor and His Institution
  (f) The Professor and Himself

The length of the Program was determined by a mixture of practical and pedagogical considerations. Although the Faculty group planning the original Program had considerable discussion about how much time would be needed to do what, it was agreed that in two months a man could write one case and also participate in other activities. Two months is not a long enough period for a man to develop proficiency in case writing or case discussion or course development. However, if a man is genuinely interested, two months does allow him to sample how the case method pattern can influence his job and the system of relationships within which he works. The Program was not like a three-day conference in which one might expect day and night peak effort. The Program took place during the summer, after a hard grading period. Participants and coaches alike needed the alternation of concentration and reflection. Therefore, Saturdays and Sundays were left free.

The closed system of eight weeks meant that every additional good idea took a lot of dovetailing or pruning off of other good ideas. Each year the proportion of time available for classroom discussion or observation as against field work, writing, and course development involved trade-offs and always differed between the beginning and end of the Program. From year to year the time available for different components changed.

All activities typically were initiated as early as possible so as to emphasize their interrelationships. This book can deal with only one subject at a time even while asking the reader to be aware of all the elements in the set of activities and relationships. Rather than endlessly repeat these interdependent interactions, this book expects that a reader will refer back or turn ahead to reports of different activities as his personal interests suggest. In this way readers may enjoy some of the stimulation which came to field research in the afternoon from discussion of a case in the morning, or the ideas about a student's sequence of learning which came from rewriting a case draft. Administration deals with complex organic processes in which literally everything is related to everything else.

An Experiment Under Special Conditions

Admittedly, no professor of business administration in any school works during the academic year under conditions quite like those prevailing during the Program. Before a participant could apply what he had gained from the experience to his activities in his home institution, he needed to be aware of the way the surrounding conditions corresponded or differed. Similarly, the reader will need to compare the conditions he faces and those of the Program. These differences do not nullify the utility of the summer experience, but emphasize *the need to explore the special conditions* under which each individual and each institution operates. The special conditions surrounding the Program are enumerated briefly as a check list to help expose and identify similarities and differences in conditions in other institutions considering the study of administration by cases.

(1) *An Institution's Objectives and Plans.* Each individual was nominated because he and his institution had some commitment to improve a specific course and curriculum. This

presupposed that the individual professor had his Dean's support for investigating and using the case method pattern. The announcement specified:

> Selection of nominees will be made on the basis of the educational plans of the nominating institutions, and the backgrounds and interests of the individuals.
>
> For a nomination, therefore, we would like a statement from the Dean of each school concerning objectives and plans for the school and how this program's experience for the nominee would fit into those plans. In this statement it is important — in view of The Ford Foundation's financial consideration — to indicate:
>
> The extent to which the school will contribute to the participant's stipend.
>
> What other support further work in case development might receive from the institution, as an indication of continuing interest.

(2) *The Participant's Commitment.* There was a corresponding commitment on the part of each participant for at least the duration of the eight weeks to explore concrete administrative situations, to write cases, to organize them into courses, to analyze cases, and to participate in case discussions sometimes as leader and sometimes as "student." Each nominee was asked for information concerning his personal and educational background, his professional interests, and the work experience he had in different industries, and a description of the course he was interested in developing. It was helpful for the coaches to have this information in arranging for company visits and other activities.

(3) *A Turn for Everyone.* This was a program of self-discovery, in which each person could make significant choices in a variety of activities, could try various alternative courses of action. It was based upon the policy of giving each participant a chance to take the roles of teacher, researcher, case writer, and student. The numerical and time implications of a turn for everyone in each of these roles were important factors in the logistics of planning the program.

(4) *A Responsive Environment — Feedback.* The fact that the participants got reactions to each of their efforts, or "feedback," is of primary importance. This responsive environ-

ment included cooperative business executives, a cross-section of colleagues in the Program, and the Harvard coaches, who were not considered "instructors" to the visitors but as "leaders from alongside."

(5) *Interaction in a Group of Peers.* A working relationship was set up within this "community of scholars" each summer. The interactions of a cross-section of men from different subject interests and from different types of educational institutions took place in the absence of evaluation from either subordinates or superiors. The participants were free to criticize and support each other, and they helped one another survive the stresses of the Program from which many individuals would have liked to withdraw at one point or another.

(6) *Concentrated Effort Made Possible by Financial Support.* From the announcement of November 14, 1964:

> The Ford Foundation has made available funds up to $800 for each participant, plus an amount equal to his first-class air travel to and from Boston. Harvard University has been asked to distribute these funds.
>
> The institution of each participant is expected to contribute a supplementary amount. This amount should bring the total received by the participant to that which he would receive if he were to teach at his own institution during the summer.
>
> The Harvard Business School, through other funds granted by The Ford Foundation, will provide discussion materials, expenses for local research travel, and typing of cases; a dormitory room; and meals or equivalent subsistence Monday through Friday of each week during the program.

(7) *Living Arrangements.* The Program initially urged dormitory living as in the Harvard Advanced Management Program. The dormitory room served the double purpose of residence and local office headquarters. The participants in the first Program advised that in the future the Program should take a "neutral" position about bringing families, as long as participants accepted fully "the rugged schedule." Some participants, therefore, did bring their families to the vicinity, or to the seashore or the mountains, and joined them on weekends. Some found apartments to sublease in the Cambridge area. But because the budget

for the Program, on which the grant from The Ford Foundation was based, assumed that men would live in one of the Harvard dormitories and take meals Monday through Friday at the Kresge Hall Dining Room, the only way to help men bring their wives and children was to make a financial allocation in lieu of the evening dinner for all participants. Many of the men also felt that it was beneficial for them to be off campus for one meal a day. Although the budget arrangements were somewhat discouraging to families, the Harvard group and their wives always enjoyed meeting those families who did come.

(8) *The Eight-Week Schedule.* Participants in the Program, at best, merely tolerated the schedule even though they understood that it was necessary to fulfill the basic policy of providing "a turn for everyone." Invariably some participants suggested that the Program be shortened, while at the same time they simultaneously urged that more activities or more time for certain activities be included. Obviously, such a closed time schedule can never be wholly satisfactory to everyone, or to anyone.

(9) *The Coaches.* An often invisible part of the Program was the participation of "the coach." His role will become both clearer and more ambiguous as each section of the Program is described. Although an essential condition of the Program, the coach is so unique in the academic world that the role is often overlooked in planning for use of the case method.

(10) *Emergent, Temporary and Expendable Community.* To reemphasize, the Program was an experiment under special conditions: a limited number of colleagues, supported by time and money and one another, trying out for a limited period the complex of relationships and activities loosely called the "case method." The special conditions discussed above were, in a sense, built into the Program as part of its basic design. The final condition varied from year to year as each group came together and developed its own identity in the interactions of the participants with one another and with the coaches from the Harvard Faculty. Very literally, "men made the difference." Those who were willing to get involved were reinforced by opportunity to improve their skill and understanding of the case method, to experience that creative leap which comes from holding both situation and subject together.

The Participants

The surprising fact is that each year an amazingly different but effective group emerged as a unique instrument by which all participants were able to experience the complex activities soon to be described in more detail. The diversity of the individuals and the variety of group patterns which emerged from pursuit of these activities in the 11 Programs suggests that the Program has some value as a simulation model by which to understand better the case method of studying administration.

Since there were so many nominations and the individual men all had appealing characteristics, selection involved first an estimate of the institutions where participation in the Program would seem to have the greatest multiplier effect. The Ford Foundation supported the Program because they also were interested in spreading acquaintance with the case method among an increasing number of institutions, while also developing enough acquaintance with the case method and capability in some schools to provide a continuing source of development.

Since the total number of 20 who could be accepted was always such a very small fraction of the total number of professors in the 600 schools and departments of business administration in the United States, selection for the Program was on a rather informal basis. Response was made to the schools that were already committed to the case method but were anxious to build a better balanced faculty. Response was made to special situations where an opportunity had opened to a school to introduce case research. The participation of men who already had a considerable amount of experience in using cases and with the practice of business, and who were curious about studying the whole process more systematically, was very helpful. Response was also made to the persistence of schools and individuals when they came back a second or third time after not having been accepted. Although one can begin to see in retrospect the emergence of "decision rules" and although the number of nominations was large enough to make the process of selection difficult, this never was such a mass operation that the "decision rules"

EXHIBIT I-2

GEOGRAPHICAL SELECTION

	1955	1956	1957	1958	1959	1960	1961	1962	1963	1964	1965	Total
Northeast												
Nominated	6	10	17	11	12	13	12	9	10	19	16	135
Selected	3	7	7	5	3	3	7	4	5	5	5	54
Midwest												
Nominated	15	17	19	14	19	24	19	17	19	20	28	211
Selected	9	6	7	7	6	11	6	7	8	9	9	85
South												
Nominated	6	6	10	11	18	9	11	8	14	11	17	121
Selected	3	2	3	7	7	3	2	5	6	4	3	45
Far West												
Nominated	4	3	5	3	8	7	8	10	12	5	16	81
Selected	3	2	2	1	3	2	5	6	3	2	3	32
Canada, Mexico												
Nominated		1	4	6	6	1	7	7	4	4	2	42
Selected		1	2		1	1	1	2	1	2		11
Total Nominated	31	37	55	45	63	54	57	51	59	59	79	590
Total Selected	18	18	21	20	20	20	21	24	23	22	20	227

were refined to the point of turning the selection over to a computer. Some of the considerations became increasingly clear over the years in trying to get a broad cross-section. The results are summarized in Exhibits I-2 through I-6.

(1) *Geographical Cross-Section.* This was in some ways the easiest index to keep in mind. Sometimes, however, in the effort to get representation from the Far West, a participant later proved to be a native of New England. Or, in the effort to get representation from the deep south, the man turned out to be a native midwesterner. There are regional variations in industry and in educational patterns, but the homogenization of the United States limits these differences. See Exhibit I-2 for the subdivisions used, and the nominations and selections over the years.

(2) *Areas of Subject Interest.* An early decision in thinking about the Program was that it should emphasize the common aspect of administrative judgment rather than any one business function. The announcement explicitly explained that many of the participants would know more about their subject field than any one of the Harvard coaching group.

EXHIBIT I-3

SELECTION BY SUBJECT AREA

No. of Participants	*General Subject Interest*
54	Marketing, advertising, sales management, retailing, international marketing
47	General management, policy, administration, small business
38	Accounting, auditing, control
25	Personnel, labor relations, psychology, social relations, industrial relations, administrative practices, human behavior
23	Finance, corporation finance, banking
16	Production, industrial management
15	Economics, foreign trade, statistics, operations research
9	Business law, real estate, transportation, insurance, business history, international business
227	Total participants

EXHIBIT I-4

TYPES OF UNIVERSITIES SELECTED

	1955	1956	1957	1958	1959	1960	1961	1962	1963	1964	1965	Total
State Universities												
Nominated	19	24	28	25	39	30	32	32	36	32	49	346
Selected	13	8	8	12	14	13	15	13	17	12	12	137
Private Universities												
Nominated	12	13	27	20	24	24	25	19	23	27	30	244
Selected	5	10	13	8	6	7	6	11	6	10	8	90
Total Nominated	31	37	55	45	63	54	57	51	59	59	79	590
Total Selected	18	18	21	20	20	20	21	24	23	22	20	227

EXHIBIT I-5

METROPOLITAN AREAS CONSIDERED

	1955	1956	1957	1958	1959	1960	1961	1962	1963	1964	1965	Total
Metropolitan												
Nominated	18	21	27	21	36	32	29	29	27	27	27	294
Selected	9	13	11	8	9	14	10	14	10	12	9	119
Non-metropolitan												
Nominated	13	16	28	24	27	22	28	22	32	32	52	296
Selected	9	5	10	12	11	6	11	10	13	10	11	108
Total Nominated	31	37	55	45	63	54	57	51	59	59	79	590
Total Selected	18	18	21	20	20	20	21	24	23	22	20	227

The cross-section in Exhibit I-3 illustrates the varied subject backgrounds which enriched discussions and highlighted the common search for points of administrative decision.

(3) *Cross-Section of Private and State Universities.* The Program also was concerned about the somewhat different problems of the large state universities and the variety of private universities. The breakdown of nominations and participants is shown in Exhibit I-4, but these categories became increasingly blurred over the 11-year period. For instance, during that time some of the universities originated by church groups received large grants from the states, or were purchased and converted into metropolitan units of state university systems.

(4) *Cross-Section of Metropolitan and Non-Metropolitan Locations.* Although this breakdown was not used explicitly, a difference in access to industry between the universities located in a metropolitan industrial setting and those located in more rural areas was a factor in selection (see Exhibit I-5).

(5) *Selection for a Balance Between Members and Non-Members of the American Association of Collegiate Schools of Business.* In 1955 the announcement of the first Program was sent to schools which were members of the AACSB. When inquiries came from a few other schools, they were considered. In 1956 announcements were sent to schools which were known to be applying for membership in the Association, as well as to members. With the approval of The Ford Foundation, announcements also were sent to other schools which had in one way or another learned about the Program and wished to be considered. Over the years the mailing list grew to include some divisions and departments of business administration which were about to be organized as schools. The grant from The Ford Foundation was for the support of "schools of business administration" rather than departments of economics and business administration in liberal arts colleges. The Program did not in any way minimize the importance of the latter group but accepted the emphasis upon a common professional orientation. Of the 116 member schools of the AACSB in 1966, 83 had participants in the Program. Of these, 8 became accredited after participation in the Program.

(6) *Selection of a Cross-Section of Schools New to the Program and of Those Extending Their Interest.* With each succeed-

EXHIBIT I-6

SELECTION FROM AMERICAN ASSOCIATION OF COLLEGIATE SCHOOLS OF BUSINESS

	1955	1956	1957	1958	1959	1960	1961	1962	1963	1964	1965	Total
Members of AACSB (at time of nomination)												
Nominated	27	34	42	35	49	46	44	33	38	33	41	422
Selected	16	15	19	18	17	18	17	19	14	13	16	182
Non-members of AACSB (at time of nomination)												
Nominated	4	3	13	10	14	8	13	18	21	26	38	168
Selected	2	3	2	2	3	2	4	5	9	9	4	45
Total Nominated	31	37	55	45	63	54	57	51	59	59	79	590
Total Selected	18	18	21	20	20	20	21	24	23	22	20	227

ing year, it became more and more perplexing whether to accept another nomination from a school which had sent able men in previous years or to turn down that nomination in favor of the first nomination of a school which had just begun to show interest. Also, in the first four years of the Program, when observation of the Harvard Middle Management Program was a central feature, the group depended less upon its own members. In later years, all benefited by including in the group men who had come from schools where they had already seen for themselves the use of the case method by some of their own colleagues and had already had the benefit of suggestions and reactions from former participants.

Later on an increasing number of nominations, both from schools becoming interested for the first time and from schools requesting additional participation, made selection more and more difficult. The Program wished to extend the number of schools and also to strengthen the nuclei of interest in schools which were already developing case-courses. In 1965, with only 20 places in the Program, 40 nominations were received from schools which had never had a participant and 39 nominations from schools which had had one or more participants. There were twice as many men as could be accepted from either new or repeating schools, and four times as many nominations in total as there were places. Clearly, no individual or school should feel that not being accepted reflects any discredit or lack of interest.

Exhibit I-7 shows the distribution of schools by the number of participants accepted over the 11-year period.

EXHIBIT I-7

DISTRIBUTION OF SCHOOLS BY NUMBER OF PARTICIPANTS

59 institutions sent one representative	= 59 participants
46 institutions sent two or three representatives	= 105 participants
14 institutions sent four or more representatives	= 63 participants
119 institutions sent a total of	= 227 participants

The extent to which professors were relying upon materials which they themselves were developing for their own courses and the extent to which they were relying upon materials developed by others was an early consideration.

There was an implicit expectation that a professor would write at least a few of his own cases each year or two as a way of testing his ideas in the field and of giving personal impact to his course. Indications of such interest were important in selecting men for the Program. With the increase in enrollments, however, it is understandable how little time many professors subsequently had available for even this limited case-course development.

When the Program began there were very few schools active in case development, so it was rare to have any participants who had any experience in case use, let alone case research. By the end of the Program, participants represented a range of cumulative institutional experience as shown in Exhibit I-8.

EXHIBIT I-8

RANGE OF PARTICIPANTS' INSTITUTIONAL EXPERIENCE

Total number of cases Contributed, ICH [a] Volumes I-X	Number of Institutions	Number of Participants in Visiting Professors Case Method Program from those Institutions
Less than 10 cases	88	150
10 and less than 20	19	45
More than 20	12	32
	119	227

[a] The Intercollegiate Case Clearing House is described in Chapter VII.

Although there was a gratifying increase in the number of schools where a significant amount of case development was taking place, it became apparent that most of the participants in the Program and the schools sending them were anticipating only token future contributions to the intercollegiate pool. For two out of three participants, about the only cases written in their schools were the ones they wrote while in the Program. It was obvious that they would probably *use* more cases than they or their colleagues would ever *write*.

Nevertheless, the Program continued to stress field work and case writing activities because these steps revealed the orientation, creative skill, and knowledge characteristic of administration.

(7) *A Mixture of Ranks and Ages.* No rank or qualification was specified in the announcement of the Program. For the first Program three deans nominated themselves, and each year there were one or two more. Opposite lines of reasoning were followed by different schools. One led to nomination of an older, well-established department head on the theory that if he endorsed the case method younger men would soon follow. Another line of reasoning led to nomination of a young man, trusting that his energy and enthusiasm would be augmented until a strong core could form. Each year an effort was made to balance the age group. In total the distribution was as follows:

68	Assistant Professors
72	Associate Professors
87	Professors, Deans, and Department Chairmen
227	

The commitment to education for business administration is well illustrated by the data assembled in the summer of 1966, an average of six years after participation in the Program (see Exhibit I-9).

EXHIBIT I-9

PERCENTAGE OF PARTICIPANTS COMMITTED TO EDUCATION FOR BUSINESS ADMINISTRATION

3%	had retired or were deceased
4%	had left education for business or government
15%	had transferred to other schools of business
78%	were continuing to serve schools from which nominated
100%	= 227 participants 1955–1965

A further factor determining selection was the coaching staff itself, a new group each summer with different background interests and qualifications. When that group was determined, it was necessary to study the nominations to try to predict which men would best help each other and their institutions to create the multiple pattern of commitment which characterizes the case method.

All these and other intangible considerations were taken into account by the Program in resolving conflicting requirements to select a group who would enjoy working closely together. The result from year to year was a mixture of Ph.D.s in Economics, Sociology, Psychology, and Statistics as well as in Marketing, Management, Finance, and Accounting. Each year there were some men with years of experience in active business but without academic degrees. Within each group the participants collectively had at one time or another taught practically every course in a business school curriculum. Each group was a representative cross-section of individual and institutional interests.

* * * * *

This book, like the Program, is directed to the active participant, not the bystander. Thus, although it is not a how-to-do-it manual, this book should be sufficiently concrete to lead the reader to a progressively better understanding of administration in business or in his own field by using the case method. In this first chapter previous efforts by those who had such interests have been sketched, as well as those who established the preconditions for the Program and governed the organization of its activities and the selection of participants.

Unlike the Program, which gave participants almost immediate feedback from many different relationships and activities, this book depends upon the reader to carry his reactions along from chapter to chapter for subsequent contrast and reconciliation. The next five chapters discuss the Program as an operating model of the case method under the following topics:

(1) Discovery of the case data in an actual administrative situation
(2) Writing the case exclusive of the case writer's opinions about the data
(3) The impact of the case data upon the conceptual structure of a course or subject
(4) Use of the case data in class with students
(5) The interaction of the case approach with the selection and evaluation of students and professors in an academic community

The next two chapters further relate the case complex to variations in the academic community. Chapter VII describes the formation and growth of the Intercollegiate Case Clearing House as an index of how widely case development and use has spread. Chapter VIII tells of the cooperative case-course developments of the Program alumni and their associates in the Southern Case Writers Association which also involves behavioral science consultants.

This book then is concerned with more than a device for instruction. It is moved by the urgent need in modern society for better administration in every enterprise. It sees the case method as an instrument for study of administration by active administrators and by professors, as well as by students. The Program it describes brought together activities and relationships too often considered separate. The focus which brought disparate views together was the search for a case — a point in the systems of economic, technical, social, and political determinants where organizational purpose and personal decision come together. At however mundane a level, it is at such points that man's discretion directs the evolution of his institutions.

Chapter II

Discovering a Case in the Field

PARTICIPANTS put a high value on their experience in the field. The Program typically scheduled visits to selected companies on the opening day or shortly thereafter. Preliminary briefing, therefore, had to be kept simple, but discussion of experiences in the search for cases went on continuously to the very end of the Program.

Plans for field work were explained to participants in the orientation session along the following lines:

> In correspondence with us, you have outlined the subjects you are teaching. The coaches have done their best to find companies in which the executives are willing to talk about concrete experiences bearing on these general subjects. In making the appointment for the first visit the coach also has explained that before any material from the company would be used it would be shown to the executive to make sure that there would be no inadvertent use of what had been given in confidence. On the first visit the chief executive may have only a few minutes to talk about his company. It is unlikely that the executive will outline a case because most executives are not thinking in those terms. Rather than be drawn into explanation of what a case is, it is generally wise to concentrate on what seems important to the executive; to demonstrate interest in his experience. As the executive sees

> that the researcher really is trying to understand him, the executive generally becomes more willing to go more deeply into critical aspects and ramifications of the situation and to introduce the researcher to others in the organization who have more information. A visit to the plant may suggest further points to explore. One may not understand immediately the relevance of what is being seen and heard. Look, listen, and record the details as soon as possible. Each coaching group will want to work together on the meaning of these data and discovery of a potential case. Obviously this approach to getting field work started makes assumptions which should be and will be up for reconsideration as the work goes on. Too much discussion in advance of all the choices involved can magnify difficulties and divert attention from the point of common interest between the manager responsible for direction of a concrete enterprise and the professor responsible for clarification of an abstract system of concepts.

Some such explanation was generally enough to get men started. Then the issues which have challenged case researchers throughout their careers could be discussed in reference to at least one finite situation. This allowed the participant to test the relevance of an abstract idea to his understanding of the things, people, and events in that case situation.

The coaches had usually set up in advance the first conference with the chief executive of a company. Even though a subordinate might have the desired data and be willing to discuss them, institutional approval of the finished case for educational purposes was a fundamental ground rule of the Program. It has proved desirable for the chief executive to meet the case researcher at the beginning and introduce him to others in the company. The researcher could then go back to the chief executive if necessary upon completion of the case. If the researcher's contacts were only subordinates, they might hesitate to release the final case on behalf of the company, and at that late stage in the process explaining the whole thing for the first time to the head of the firm might prove embarrassing.

During the summer many members of the Harvard Business School Faculty met with participants to discuss their experiences in case-course development. Participants likewise benefitted from a great variety of views about field work.

The Relationship of Case Researcher to the Executive

Coaches followed different methods of working with participants in starting field work. Although many followed the practice of an initial group interview, the typical pattern was to let the individual participant follow up on the case lead which most appealed to him. One group which continued to visit companies as a group made a practice upon return from the field to sit down in the coach's office and dictate into a recorder the impressions received from different members of the group. The participants thus helped each other to distinguish between the data which could be observed by others, and impressions that an interviewer received from his own questions, and the effects of the exchange upon the relationship with the interviewee.

Participants soon saw that this case research relationship is unique and difficult to describe. The company executives were also trying to define it. Even when a participant entered an organization "through channels" and felt he had achieved a reasonably good understanding of his project at the top, he still found it necessary to reintroduce himself and his mission to each member of the organization from whom he sought information.

There is no easy analogy. The case researcher-informant is *not* in the relationship of doctor-patient, consultant-client, FBI-witness, teacher-student, superior-subordinate, or news reporter-celebrity. While acknowledging the importance of clarifying this relationship, coaches warned against diverting the interview by elaborate explanation. Rather, they urged participants to *demonstrate the case research relationship* by manifesting an unmistakable singleness of purpose in learning what was important to the informant about the case situation being studied. Building the relationship with executives is the foundation of all case-course development. Even the most seasoned researcher often has perplexing moments and must create this relationship anew in each new inquiry.

Apprehension in the first few days of each Program, about getting executives to talk and what "quid pro quo" might be expected, gave way to concern about how deep a commitment the researcher was willing to make because an executive frequently warmed to an attentive listener. The professor-researcher found that the situation in which he had only a passing interest for an hour's discussion in class might be a crisis in the career of his informant and his company. The question then turned to how in good faith an interviewer might back out after his initial protestations of desire to understand this concrete situation.

In a seminar discussion of field work, participants found it thought-provoking to consider their research relationship as a "psycho-social contract." This was discussed in several different Programs by Harvard Business School Professor Abraham Zaleznik and one of his former Research Associates, Professor Richard Hodgson of the University of Western Ontario, who had been studying role constellations of the executives of a mental hospital.[1] Some of the points which caught the attention of Program participants are paraphrased below.

> Research methodology is a function of the researcher's concern to get authentic data in depth.
>
> The relationship of the researcher and his informant is based on a "psycho-social contract" parallel to *but different from* the doctor-patient or consultant-client relationship. It is more implicit than explicit.
>
> The researcher brings prestige. The business executive has power. The exchange should be in terms of learning-feedback.
>
> The expectation must be embedded in the researcher's own behavior. He does need a definition of his research interests, the intellectual reason he is there.
>
> He has an active role, based on curiosity. The document is the product of his thinking: "my work." The release indicates that the executive also has been learning something and his signature indicates his approval of the learning relationship.
>
> The reciprocity in the research relationship is characterized by the researcher's own behavior as "giving attention" rather than "taking notes."
>
> The way to do research, is to do it. Work in the field until uniformities appear, emerging from the data themselves.

Professors Zaleznik and Hodgson urged participants to take

[1] *The Executive Role Constellation: An Analysis of Personality and Role Relations in Management,* by Richard C. Hodgson, Daniel J. Levinson and Abraham Zaleznik (Boston: Division of Research, Graduate School of Business Administration, Harvard University, 1965).

responsibility for exploring a creative role significantly different from that of either the prosecuting attorney or the fly on the wall. Since there was no widely known model for this role, each case researcher would have to establish it for himself in each case situation, by his behavior as much as by anything he said.

Organizing Ideas and Order in the Situation

The collection of data is also influenced by the ideas the researcher has of the enterprise and the subject being studied. The Program was enriched each summer by seminars with scholars and men of affairs who stimulated discussion by their theories of organization, science, art, and administration. (See Appendix 3.)

If the company is a going concern whether or not a professor is there to observe it, what is so difficult in going out to see it? Real men and women buy for it, use its machines, sell for it, grow old in it. It exists as a system in its environment. But the researcher is challenged because he cannot see it directly or all at once. Men by the hundreds or thousands may have a sense of belonging to *it*, and *it* is the continuing reference against which the administrator is pondering his choice of direction. But none of those who "belong," not even the chief executive, can show it to the researcher completely or pull out a photograph of the living reality. Yet this is what exists when the researcher is not there, and what the researcher is trying *to discover* in relation to his subject, somewhat as the administrator himself is ever rediscovering its direction of growth under constantly changing conditions.

Participants soon realized that what was going on in their relationship with executives and the data they were collecting had something to do with the ideas they were taking into the situation. In discussing these phenomena, the Program was privileged for several years to have the point of view of Professor Renato Tagiuri of the Harvard Department of Social Relations. (He subsequently was appointed Professor of Social Sciences in Business Administration at the Harvard Business School.)

As with any other method, the case method has many arbitrary aspects. Certainly one such arbitrary aspect is due to the particular orientation of the researcher or case writer. *What will be observed in the field will depend on the case writer's own interests and background.* Every human being has a more or less arbitrary framework in terms of which he orients himself in his environment, *a kind of implicit theory* that helps him to make decisions at every moment of his existence. Furthermore, the kind of observation he makes will depend on the framework that he develops while in the situation.

. . . The important thing about studying a situation by the case method is *the attitude of the investigator.* The useful attitude is present when the student enters a situation self-conscious of the conceptual scheme he carries with him. He is then in a fairly good position to see for himself whether his scheme is a useful one for making observations. If, for instance, he *cannot* fit the situation to his conceptual scheme when he is applying it in all honesty, then he is bound to learn something. It may be that some of his concepts are not sufficient to cover certain circumstances or that others are too broad and therefore do not discriminate instances among which, in the light of other evidence, it seems desirable to distinguish.

Just how the ideas taken into a situation influence what one sees and hears is far from simple, but by the end of the Program, many participants noted that increased skill in observation and interviewing was in large measure a function of increased awareness of one's own feelings and beliefs and their projection into the situation. The expert knowledge some professors had in specific industries or business functions, and the specialized backgrounds of others in mathematics, statistics, or social science, did not assure insight into *what the facts in the specific situation meant.* The connection between knowledge one brings to a situation and the "order" already in the situation is far from simple. Even the possibility that there might be a difference between the two, and that recognition of the difference might help one learn, stimulated discussion.

Participants typically were surprised to find how much work with the data the coaches expected. Apparently before attending the Program, participants had the stereotype of the ideal case researcher as a sort of human tape recorder whose notes somehow became a case without effort. Frequently a professor attending the Program would proudly return from an afternoon at a company and throw his notes down on the coach's desk. "Here is the case. Exactly what the executive said." It took a great deal of tact for the coach to get the researcher and other members of the team to consider why the executive talked about some points but was completely silent on others which the participants had thought important as they took their initial tour of the plant.

Occasionally cases which have become "classics" came, it is true, almost verbatim as a researcher made notes of an anecdote. Surely one should not spurn these windfalls when they occur. Typically, however, the researcher gains understanding of the situation only after careful accumulation of data and working with them until they begin to fall into place.

Although at the start of the Program it was stressed that the researcher should focus on what the executive thought was important, that did not mean that the case came without further struggle for and with the data. The early emphasis on structuring the relationship with the executive and others in the company was for the purpose of helping the researcher get direct access to *what people do, with whom they do it, under what conditions,* and *what these activities, interactions, and conditions mean to those involved.*

These categories are cogently clarified as elements in a dynamic social system by George C. Homans in *The Human Group.* In that book Professor Homans also illustrates the challenge that guest speakers presented to Program participants by the opportunity to interrelate what was being found in concrete field case situations with abstract models. Many case researchers, for instance, were involved in situations in which executives were anxiously trying to create new "customers." Professor Homans asks the general question, "what makes custom customary."

> Men cannot plan for the future without relying on the massive regularities of expected behavior. Yet when everything intelligent has been said about the usefulness of custom, one more profound question remains: What makes custom customary? For the brute fact is that customs do change. In view of the constantly varied forces playing on society, it is amazing that anything can be recognized as persistent. The recurrences are miracles, not commonplaces; and miracles, if they happen often,

> are just the things we should study most closely. As soon as we do, we find that nothing is more defenseless than a custom, alone. Not single customs, but systems of custom, survive. Anthropologists used to talk about the "tyranny of custom" as if custom were a mold pressing social organization into a shape. This view is misleading. Custom is not something outside of, and apart from, social organization but is implicit in organization.[2]

Getting the Data

If the data of administration include what people do and their interactions with others in doing it, then the researcher needs to consider what he can learn by observation. These data are not solely in words. Participants were sometimes irritated as well as stimulated by seminars conducted by Edward T. Hall.[3] He explained to participants that such nonverbal communication often takes place out of one's awareness. Dr. Hall continued that professors ordinarily have been so immersed in the world of words that they fail to read the nonverbal data from which the practicing administrator corrects the "model" of his firm from day to day.

Professor Hall was stimulating because he multiplied the means of inquiry, without risk to relations within the enterprise, by increasing the researcher's alertness to the nonverbal systems of communication. A direct question may actually block access to the organization, or if raised prematurely make an informant suspicious. Inquiry does not necessarily mean asking questions, although it is imperative to have them in mind.

Participants during their initial plant tour and other case visits tried to use Hall's clues to communications from the use of space and time in the company. In discussions later, participants differed greatly in their capacity to relate what they saw on these visits to practical business problems; for instance, the com-

---

[2] *The Human Group*, by George C. Homans (New York: Harcourt, Brace and Company, 1950), pp. 28–29.

[3] *The Silent Language*, by Edward T. Hall (Greenwich, Conn.: Fawcett Publications, Inc., from Doubleday and Company, Inc., 1959); and *The Hidden Dimension*, by Edward T. Hall (New York: Doubleday and Company, 1966).

plexity and extent of capital equipment in relation to the work force and inventory of raw materials, work in process, and finished goods. Similarly, it took considerable discussion within the team before participants saw what these physical factors meant to the financial position of the firm and potential pressures on executives and work groups, and how these factors related to the intangible human situation. It was helpful to the members of the Program to see the company's finished product in the marketplace. With a little thought about these gross observations, the participants were able to understand better the system in which the administrator must exercise discretion. But for some cases obviously more careful observations were needed to discriminate various attitudes.

Although activities and interactions may be observed, attention frequently was preoccupied with interviewing. Therefore, much discussion of field work in the Program centered on what the researcher was trying to learn from his interviewing. Often it was seen that the researcher was trying to learn about work flow and organization. Clues to the values and the meanings which these activities and interactions held for individuals and groups sometimes were not recorded. Some of the coaches, therefore, stressed the immediate recording of interview notes to get the characteristic expressions which might give an index to the values which were important to various people in the enterprise. Expressions of likes or dislikes were to be particularly noted. These discussions brought out that working with the data meant wondering *why* a statement was made, rather than whether or not it was "true" or "false." [4]

Listening as a way to get data seemed obvious because introduction into an administrative situation immediately plunged the researcher into an interview. The case interview is not a cross examination. Often, by analyzing the data, the times and places of events will become clearer. Also, details of technical and economic characteristics may be more accessible from documents than from one man's memory and surely less time-consuming.

[4] Participants were given copies of "How Do You Know If the Informant is Telling the Truth?" by John P. Dean and William Foote Whyte (*Human Organization*, Vol. 17, No. 2, Spring 1958).

Experienced researchers pointed out in other seminars how executives lost interest if their valuable time was being used by the researcher to get information he could have obtained by his own "homework." In contrast, researchers told of busy executives who canceled other appointments to keep on talking when the interview had become focused on the *meaning* of things, people, and events in a puzzling situation and the mixed feelings that the executive was struggling to clarify.

The classic rules for interviewing were discussed in most sessions by many researchers in various areas of business administration. The following statement of them is taken from Elton Mayo's chapter on "Psychopathology and Social Study," [5] in which he was reflecting upon the framework underlying industrial research:

> . . . But he who seeks to understand and live contentedly with his fellows must begin from where their thinking actually is and not from where the textbooks say it should be.
>
> . . . Look for the significance of all statements in the situation of the person that makes them. Consequently the simple rules for the conduct of any clinical interview, whether medical or sociological, are as follows:
>
> (1) Give your whole attention to the person interviewed, and make it evident that you are doing so.
> (2) Listen — don't talk.
> (3) Never argue, never give advice.
> (4) Listen to:
>   (a) What he wants to say;
>   (b) What he does not want to say;
>   (c) What he cannot say without help.
> (5) As you listen, plot out tentatively and for subsequent correction the pattern that is being set before you. To test this, occasionally summarize what has been said and present it for comment. Always do this with caution — that is, abbreviate and clarify but do not add or "twist."
> (6) Remember that everything said must be considered a personal confidence and not divulged to anyone. (This does not prevent discussion of a situation between professional colleagues. Nor does it prevent some form of public report when due precaution has been taken.)

---

[5] *Some Notes on the Psychology of Pierre Janet*, by Elton Mayo (Cambridge, Mass.: Harvard University Press, 1948), pp. 21–23.

The value or meaning of events to a person is a major reason for interviewing him. Therefore, in the interview, the interviewee's characteristic words and ways of thinking are the data being sought.

The relationship between field work and work with the results is very intimate and intricate. Often the obvious was overlooked by the man starting on case research. It took repeated exposure to a situation to find that the activities and relationships in the situation were interdependent, that analysis of data was part of the preparation for field work and vice versa. Once the case researcher was set up in relationship with an active executive and had become involved in the use of cases in class, he found answers to his questions literally by going back and forth between the field and the discussion group. The continuing connecting thread as in all successful research was curiosity, wonder, and the search for congruity. Getting the data was only a fraction of the "field work" job. Experienced case researchers reported a great variety of ways to build the basic record of what they saw and heard, and exhibited different working papers they found helpful in analysis.

Working with the Data

One set of working papers would usually show the physical layout of factory and office, and the sequence of operations. The idea of the flow of work is not confined to operations in a factory. It would apply also to steps in securing raw materials and the physical distribution of goods from a factory to the ultimate consumer. These operations are the economic justification for the enterprise. Coaches would try to get the participants to visualize as clearly as possible the physical activities that must be accomplished and the environmental and other conditions which affect these activities. The wide range of interests in each Program group helped the group as a whole to note the informal activities which might otherwise appear to have no utility to the professor trained as an economist or engineer, for instance, but which to the actual organization might be immensely significant.

The formal organization chart was another work sheet which

frequently brought to light differences between observed relationships and what some of these relationships "should" be according to the top executive's statement or the researcher's organization theory. The coach and several participants who had visited a firm were able to compare more fully the relationships which arise from the spontaneous interests, likes, and dislikes of men and groups, and they brought to the attention of the other group members activities or relationships which might have been dismissed as "irrelevant" from only one point of view.

Another work sheet to help understand the enterprise as a living whole was a chronology of events. This enabled the researcher to compare the present issue with the way the enterprise had dealt with situations in the past. The purpose was to get a sense of the direction or life flow of this enterprise and the impact upon it of alternative choices which confront management at a given time.

Working with the data being collected proceeded simultaneously with obtaining data. The coaches encouraged an analysis of these data after each visit. Researchers were urged repeatedly to look and to listen in the field, to concentrate on what was important to the executive while they were with him, and not to influence him with their own ideas. But when they returned to the School, the coach urged the researchers to shift roles and, after recording the data, to make their own independent analysis. The researchers, of course, were encouraged to record any size-up which the executive or consultants had made, and to compare this with their own analysis.

Understanding the "system," however, does not necessarily lead one to recognize *a case.* It was only by working with the data, and by more discussion within the coaching group that the idea of the case itself emerged.

The Case as the End and Means of Discovery

Paradoxically the case is what the researcher wants to discover, and the idea of the case is the means by which the researcher ultimately completes his inquiry. The idea of the decision case establishes the *boundaries* around the data. At the same time, the relevant data calling for decision also create the case. This relationship between a body of data and the point of decision, between a system of determinants and the point of discretion, was a continuing topic of discussion.

Terms in which the case method was frequently discussed added to the confusion because they often implied a distortion as well as a partial description of the process in which the participants were engaged.

> "Case history," although useful, is different from the search for future purpose.
>
> "Case collection," as has been seen, understates the work the researcher must do in finding the data and their meaning.
>
> "Case writing," as will be seen in the next chapter, is a significant literary effort but the term understates the painstaking inquiry for stubborn facts and their interrelationships.
>
> "Case research," which we have used so often, implies a repetition and subordination of the *search* into the unknown future which is such a critical feature of administration.

Hence *the case as the thing* to be discovered in terms of concrete things, people, and events, and *the case as the idea* of an issue requiring action created by a set of facts in a concrete situation, alternately moved in and out of focus. Once the distinction was realized, the semantic problem lost interest. But until the reality of the idea and its existential projection had been experienced, efforts at explanation often seemed to the participants willfully frustrating.

What the participants were looking for in the field was first-hand experience with a body of data involving an issue of action, the need to reinterpret purpose. At least this was what the coaches were trying to arrange. One cannot infer from the written case what the situation was like before the researcher became involved in it. Part of the paradox, though, was that although a case might be in the situation, much to the surprise of participants the case often was not recognized as such within the company. It is by involvement in the situation that the researcher, from within it, *discovers* the cluster of data creating an issue in the system requiring discretion, a choice of purpose.

Technically the case could be thought of as a way to find

relevant boundaries around an organic body of data. The coach and the researcher might become *aware* of an issue which was implicit in the data, an issue of which the participants in the company were *unaware.* The researcher might then need to determine what further data would be needed to make a decision on policy direction and to get at least an approximation of these data into the case. Reviewing first efforts by participants revealed the shallowness which comes from the researcher's settling for data superficially suggested by the logic of an issue he had brought into the company without having projected himself into the living relationships of the particular situation.

In discussing "the case method culture" with participants, Professor Edward T. Hall pointed out that anthropologists typically work either at the level of pots or of populations. The case presents a different scope of inquiry. The "construct" of the issue of action faced by the administrator, as a way of fixing the level of data for a case, deals ingeniously with "the principle of indeterminacy in culture." "Isolates turn into sets when they are studied in detail and are therefore abstractions. The more precise the observer is on one level, the less precise he will be on any other. Only one level can be studied with precision at any one time, and only one level can be described at one time." [6]

When a man discovered his own case, it was an exciting part of the Program, but this process of discovery did not seem quite respectable by some academic standards. The decade of the Program was also the decade behavioral scientists and business schools were actively exploring mutual interests and methods. The idea of "the case" and the reaction of social scientists who came to speak to participants stimulated numerous fruitful discussions. One of the provocative speakers in this exploration was Professor Raymond Bauer of the Harvard Department of Social Relations (he was later appointed Professor of Business Administration, Harvard Business School). Professor Bauer made the following point:

Somehow there has developed a myth that there are two kinds

[6] *The Silent Language*, by Edward T. Hall (Greenwich, Conn.: Fawcett Publications, Inc., from Doubleday and Company, Inc., 1959), p. 170.

> of research: "case research" and "scientific research," and that these two are in opposition. I want to take a position of firm opposition to this dichotomy. It is true that "case research" would generally be used at the "exploratory" or "hypothesis generating" end of the continuum rather than at the hypothesis testing end, although case (or clinical) research can also be used for hypothesis testing. It is further true, as expressed above, that most discussions of research have taken as the paradigm of good research the hypothesis testing end of the continuum. However, the generating of hypotheses and their formulation in testable *and significant* form is an essential part of the business of research and one which is just as vital as the testing of hypotheses. Furthermore, there are various ways of testing hypotheses and the crucial experiment is only one of them. Many of the most important propositions of any science are not amenable to direct test but undergo the test of attrition in use.

Presentations by natural and social scientists in Program seminars helped to identify the unique feature of the case which the researchers were struggling to find in the data of their field work. It slowly became evident that the data peculiar to administration include the ever-shifting points at which discretion must be exercised within or at the intersection of different "systems." Various scientists will abstract out different systems or subsystems from the totality of the enterprise and its environment. What they learn is of importance to the administrator, but it misses the essence of administration, which is to supply the "givens" which the scientist must assume in order to push forward his controlled experiments. It proved useful, therefore, to explain the case as this moment of intersection between the determinants studied in isolation by the scientist and the need of the enterprise for the exercise of discretion by the administrator. The knowledge of systems in the behavior of men and organizations does not automatically tell the administrator what direction to take. There is a point of action where a person facing responsibility may realize he is in a network of determinants, but he is aware also that he has to exercise discretion. The case situation involves this mixture of determinants in a system and of the discretionary possibilities that are open at the point of decision.

The "old pros" of case research, including Dr. Melvin T. Copeland, George Fisher Baker Professor of Business Administration Emeritus, author of the first case book, generously met with the Program year after year to consider the conditions for successful investigation. Over and over, their reports confirmed the experience of the Program participants: From patient exploration of relationships, the researcher eventually realizes with some surprise that he "knows" what circumstances determine the situation *and* where there remains a seemingly small but strategic point at which the responsible administrator can exercise discretion. This discovery is not deduced from the conceptual scheme, although some concepts are more useful than others for directing inquiry.[7] It is akin to the process called "Autonomous Discovery" which Dr. J. Richard Suchman, one of the outside speakers, analyzed as "psychologically, *an imaginative leap* in contrast to rigorous logic." Both mental processes are needed, but in the creative process of administration the capacity for discovery has a premium and the discovery of the case itself is a thought-provoking analogue.

Not every "imaginative leap" turns out to be a significant discovery. Much effort was devoted to analysis of data and the testing of proposed actions. Coaches were concerned with participants who jumped to conclusions and were not willing to test them. "Do you have the data, as well as the impressions, from which others may make the discovery for themselves?" Are there data from which to consider alternative courses of action?" "Do you have the data which your students will need for rigorous analysis?" This is the point at which the researcher needs the expert knowledge of his subject and skill in developing hypotheses from his understanding of the situation.

The researcher who needs additional data cannot always obtain what he would like. He is still a guest in the company and must feel his way along within the relationship he has established there. Since case material often is disguised to eliminate any identification of the company under examination, the researcher may need to devise indexes to show relative magnitudes instead

[7] Comments on this direction of inquiry by Associate Dean Lombard are presented in Chapter V.

of using specific company data. Maintenance of basic relationships in this way while disguising details can realistically convey the issue calling for decision. The reaction of the administrator when the finished case is submitted for his approval tests the researcher's maintenance of confidence and his research relationship. When the search for understanding has been mutually beneficial, the executive often goes to considerable trouble to help state data in ways which capture the essence of the situation even when disguised.

Establishing the Case Lead

Some such idea of this process of case discovery was in the mind of the coach as he searched before the Program started for leads to companies where his team might find cases. Since participants ordinarily were not directly involved in this search, and since this function of the case hypothesis is so important to fruitful case research, the method will be described further.

The business executive obviously knew much more about his enterprise than the coach. The professor of a particular subject probably knew more about his area of special interest than the coach. Therefore, the role of the coach or, in his absence, of the researcher, is to help find the moment where the knowledge of the professor is relevant to the company's need for action.

One lead for possible case development was a large company which had recently acquired a manufacturer of knit wear. In talking to the chief executives of this company, together with a professor of marketing, the coach found that both the company and the professor were interested in discussing credit as a means of sales promotion. The professor later confessed that he had not seen any case possibility in what had been said and was surprised when the coach had asked the business executive if he had some specific accounts where the sales manager and the credit manager had differences of opinion about accepting orders. This question prompted a vigorous response from the manager. He told of a salesman (returning from his territory) bubbling over with enthusiasm for five new accounts, only to find that the credit manager vetoed orders from all five because they did not meet credit standards. Arrangements were made for the professor to meet with the credit manager and sales manager, from whose conflicting interests he subsequently found an excellent case. In this instance, the expert in marketing and the chief executive, although having an interesting conversation, were about at the end of meaningful discussion with each other. The coach, having in mind the case idea, picked up an "obvious" point of action at which the conceptual framework of the professor-expert could be relevant to the needs of a business executive for policy direction. Once this action point was seen, the discussion was focused and plans were made to explore specific conditions.

In another instance, a professor and coach were calling on the manager of a printing company to explore experience in working capital management. The business manager casually mentioned his consideration of a bid on a very large printing of a long book. The lapse of time between the purchase of paper and receipt of payment for the job would be vastly different from the time element involved in his usual small jobs. The coach saw a case lead around the impact of the unusually large order on the shop operating costs and working capital, not because he knew more about either the flow of funds or printing, but because he was on the alert for some concrete point in the life of the enterprise where the researcher's system of knowledge might make a difference in exploring the situation and in the manager's decision.

The Program sought to multiply exposure to this *process of case identification* by scheduling group visits in the early phases of the Program. Thus, in many of the Programs, a coach would take from three to six members of his team to call upon several companies. In this first orientation visit, the hope was to have a general conference with business executives as well as a tour of the plant. Then the team in discussing their experience would discover which member seemed to find the company most interesting. This plan of group initial visits worked surprisingly well, considering that the Program needed 15 to 20 or 25 companies that were ready, willing, and able to receive visitors within a week or ten days from the beginning of the Program. One can see why the ideal situation a participant might have specified was not always found. But this is normal. In all research there must be trial and error. There were a few embarrassing times when none of the participants turned out to be interested enough in a company to follow through on getting the data and writing the case. The coaches subsequently became more circumspect in promising an executive that a case would be written about his situation.

The few instances when the lead turned out to be disappointing provided opportunity for discussion of this very probable occurrence. Does one modify his expectation and write up what he does find? Or does he cut his losses by withdrawal as politely as possible to search for another company in which to test his concepts? Where a group was visiting four or five companies by prearrangement of the coach, the participants had little practical alternative within the time constraints of the Program but to pursue one of the leads which had been identified. There were exceptions, such as a visit by one team to a Philadelphia company that had extended an earlier invitation to one of the participants. Occasionally other participants, after seeing what possibilities were available in the Boston area, brought out material they had gathered in advance from a contact at home. Because of the Program's time pressures and the fact that several companies were made available through prearranged planning, many participants wondered what they might face when seeking a case on their own.

In the first year of the Program, one of the coaches did wait until his team arrived in order to discuss with them how they would go about finding a company with a mutual interest and a situation that would lead to a good case. This process of searching out a company situation in which there was a high probability of common interest was an extremely valuable experience. However, it turned out that eight weeks was much too short a time to search for the case lead and then follow through with finishing the case. Thus, in subsequent years, armed with correspondence about the participants' interests, the coaches endeavored to have several case leads established before visitors arrived. To make up for the pre-arrival step, coaches discussed with participants how the lead for each company had been developed. The search for a case lead also is a challenge to the relevance of the theory one is testing. Discussion of case leads therefore becomes a discussion of course objectives and concepts.

Discussion of establishing case leads brought reactions at both extremes: anxiety about getting access to any company and skepticism that *following* leads could develop one's *leadership* in the profession. Both views may miss the mutuality, the creative interaction between the professor and the practitioner being sought. The anxious one may overcommit himself to completing a case in any company where he gets an introduction regardless of the needs of his course. The skeptical one may lose perspective in his advanced ideas if he does not accept the challenge to test their relevance in the realities of a concrete situation. Over the summer, therefore, the Program invited many experienced case researchers to describe their practices in selecting and establishing case leads.

Typically, a professor developing a course from cases continually scans the business world for the case situations he needs to test his hypotheses. He is always on the alert for clues to help him find an executive facing a situation in which his subject ideas may be tested. He keeps an ever-expanding file of case leads. The leads he can develop in his own community are not constrained by the eight weeks of any Program. In the continuing relationship which a professor can develop with business executives in his own community, also, he has more possibility of finding "the hot lead" or of waiting until the situation in a particular company is ripe for executive attention.

In the first stages of developing a case-course, there is the advantage that almost any case situation provides a significant building block. In a subject in which case-course development has progressed further, the case researcher probably needs to be more selective in his efforts in order to fill the blank spots in his course. By that time he will have developed a much wider acquaintance and acceptance in the business community and will have had time to build up a file of potential case leads. Then he may well follow the plan of making a survey by brief interviews before settling on the particular enterprise in which to study the case situation in depth.

Professors experienced with building courses from cases told of making such a preliminary survey of several potential leads. They would explain their interest in a subject and get a brief statement from the company about its current experience with the subject, without any commitment to study the company in depth. After a vestibule study of three or four companies, the most promising one would be approached for a detailed study. For courses which have been developed to the point of such selectivity, this approach multiplies the background of the professor and may turn up a company which not only faces the immediate issue to be studied but does this in a context which contributes effectively to other needs of the course.

Professors who had pioneered new courses demonstrated that case-course development is a point of view. They identified cases which had become classics through repeated use, and told how they had been recognized in anecdotes heard in Pullman or airline conversations. The need to make decisions is so much a part of the executive's life that case vignettes are almost unavoidable in his conversation.

Professor Malcolm P. McNair, for instance, explained to participants in a seminar discussion that the present practice of submitting cases to companies for "release" came about because he had used a smoking car anecdote as an examination. This was subsequently published in the School's catalogue to illustrate the type of situation used as an examination. A friend's son called it to the executive's attention. In outrage, the executive phoned and complained, "McNair, that was a private conversation. I had no idea you were going to use it in class." Thereafter, the Faculty of the Harvard Business School made sure that executives had a chance to see what was going to be used so that there would be no chance of unintentional breach of confidence. This is only one of numerous cases that illustrate what may be found by alertness to various decision-making situations.

Although men new to case research often become unnecessarily anxious about finding case leads, once the process has actually begun, the case researcher is often embarrassed by having more case contacts than he can pursue promptly.

Division of Labor in Case Research: Role of the Coach as an Illustrative Possibility

Discussions of field work in the Program made it obvious that the researcher was very much *involved* personally. It was less clear that what he discovered and how he tested the results depended upon *awareness* of this involvement. Although the line between "enough" and "too much" help in clarifying the involvement preoccupied the coaches, the need for division of labor and teamwork inherent in case development was not adequately realized. It may help to understand the field work experience, and other activities in the case method set, to think in terms of how one might deal with the division of labor between the researcher and the coach — even if one were alone and playing the two roles alternately.

Most cases involve some sort of cooperation in authorship. Among the 3,000 most widely used cases in the Intercollegiate Case Clearing House at the end of 1966, some form of collaborative authorship was involved in:

7 out of 10 cases in Marketing, Business Policy, and other General Management;
6 out of 10 cases in Control and Finance; and
5 out of 10 cases in Production and Human Aspects.

The division of labor may take many forms. The listing of the 3,000 most widely used cases showed, for example, that the tasks might be shared by a student, a research assistant, an editor, a colleague, a consultant, or a professional group meeting to examine drafts of each other's cases. The deliberate effort by the coaches during the Program to remain unobtrusive makes the role all the more difficult to describe. Maybe that is a fundamental quality of the relationship. The most frequent pairing of functions has been between the instructor developing his course and a present or former student serving as his research assistant, and this also is not an easy relationship to establish. For three years, 1953–1955, the Faculty Committee on Case Development Policy of Harvard met with small groups of instructors and research assistants to see what conditions made this relationship productive or unproductive. The findings were in part:

> . . . In the relations between a senior Faculty supervisor and a junior Faculty man assigned to case writing . . . it helps the work when the senior, (a) is going through a process of re-examining or rethinking parts of his course at the time he is supervising the junior, (b) if he shares his course planning ideas with the junior as they plan their case collecting efforts, (c) does some of the field work along with the junior, (d) has the junior's office near his so that the necessary frequent contacts are facilitated, and, most importantly, (e) feels some responsibility for the balanced development of the junior's capacities as well as a sense of responsibility for producing useful cases.
>
> . . . As a Faculty we apparently need to remind ourselves on occasion that there is no "ideal" case type apart from particular teaching objectives so that we do not build artificial barriers to the attainment of those objectives.
>
> . . . Some people indicate more by their actions than their words that they think of good case writing as an essentially simple, routine, unimaginative "hack" job. . . .
>
> . . . The job of the research assistant has the potential of being highly rewarding . . . , but the senior Faculty must assume an increased responsibility in seeking out these people early in the school year and constructively presenting the possibilities of the job. . . . Experience indicates that this is especially important with some DBA candidates who may be preoccupied with thesis interests to the detriment of the production of teaching cases. . . . The interview results indicate that the senior Faculty is well aware that it takes considerable time to do a good job of supervising a junior on case writing. Sometimes our practices on this matter do not conform to our intentions. . . .[8]

This report was being considered by the Harvard Faculty at the same time that the Program was being designed, and every effort was made in setting up the job of coach to avoid the impression that the visiting professor was the assistant or that the coach was the instructor. Considerations of status may have diverted attention from analysis of the division of labor, which could be made in many different ways, in order to increase awareness of the complex involvement in the business

---

[8] Report to the Faculty from the Case Development Policy Committee, December 16, 1955 (Mimeographed), pp. 8–12.

enterprise and in the academic community required for case-course development.

The Program underlined the fact that one cannot study a situation without getting into it. How self-conscious one need be of his influence upon what he is studying may be debated, but nevertheless the investigator is *in* the situation, or dependent upon the presentation of data by someone else involved in it. Some coaches felt that all one needed to do was to tell the case researcher, "Go out and get the facts." Often this was enough, but also strange things were sometimes brought back. At one extreme, the case researcher came back with a verbatim transcript of only what the business manager told him. "This is his story." "I kept my mouth shut." At the other extreme, a case researcher came back with what he had when he started out, prejudices only now confirmed by polite answers to his leading questions. Needless to say, both of these were unproductive extremes and would sorely test teamwork under any division of labor. Awareness or lack of awareness of what is going on in a situation has many perplexing ramifications. The case researcher with little or no field experience, who may have read many cases, often expects that the executive he interviews should have some such "case" ready made in his top drawer, or at least clearly in mind. The researcher is surprised, therefore, to find that often no one in a company seems to know what is going on — that he, the researcher, is learning things of which no one in authority seems to be aware. In the Program the coach was at hand to discuss next steps. Frequently in the Program the coach and the rest of his team had some independent view of a situation because they had made the first call on the company together. Thus, they could check impressions of their colleague as the work went on. The objective was to juxtapose the concepts taken to the situation and the data of the situation with minimum distortion and maximum illumination. Even a brief firsthand view of the facilities and conference with management gave the coach and other members of the team clues which helped them to test the subsequent reports of the researcher who was concentrating on that enterprise. The process of enlarging awareness, however, does not depend solely upon an independent view of the field situation.

A business enterprise is both persistent and ephemeral. The observer gains almost imperceptibly a sense of knowing the enterprise. This capacity to find the significant elements in situations is an objective for students which the case is to strengthen. This sense of acquaintance must be obtained and communicated by the ephemeral bits of observation and interview data which lend concreteness to the description.

The division of labor in case research and the consultation involved does not require formal expertise as much as an awareness of the process and the value of different points of view. When case research was first organized at the Harvard Business School in the early 1920s, each group of three new case writers had a "section boss," who often had only a few months more experience than his team. The initial responsibility of looking at the process proved to be of lasting benefit to such pioneers as Donald K. David, Malcolm P. McNair, Neil H. Borden, and others who served as section bosses and then went on to build careers in the study of business administration from this foundation of getting and helping others to get the basic data of administration for cases. In finding leads, for instance, there was and is the need to match the requirements of a course, the personal interests of the researcher, and the probable point of action in a particular enterprise. When the researcher is introduced to the company, both institutional expectations and personal interests are involved. As already pointed out, the conduct of the research depends upon the researcher's demonstration of his interest in what is important to his informants. This very involvement, however, makes it useful to have a coach or a colleague close at hand to discuss the original purpose and to provide the detachment needed for a balanced selection of data. Equally important is the challenge to pet theories in a course which is presented by a statement of the facts by an independent observer. Finally, at the time of returning to the company for release of the case, the coach or someone in his place is a valuable resource in the event negotiation or reassurance is needed.

In the Program, search for a case in the concrete experience of a company was reinforced or redirected by concurrent discussion of cases in the classroom and by efforts to sequence these and other cases in the design of a course. Likewise, classroom discussion and course design were illuminated by the complex relationships emerging in the field.

Many elements of the field experience merit further development, but as in the Program itself, this will be done in considering what aspects of the experience go into the written case, what aspects go into the sequencing of this case with other cases and concepts in design of a course, and what aspects of the total experience may be recreated in classroom discussion. The task would be simplified if the pieces could be isolated, but their interdependence is the essence to be communicated.

Chapter III

The Written Case

The preceding chapter dealt with the case situation in the field and the learning involved in its discovery by the researcher and the administrator. The task now turns to recreating something of this learning experience and multiplying it for large numbers of students.

The case situation in the field is more than just words, but somehow words must serve as a bridge to the academic world. In the classroom and elsewhere in the curriculum the company's life experience can be reconstituted by emotional involvement and nonverbal behavior, but for study purposes the blank pad of paper needs words. Not just any words, though, because the finished product must be presented to the informants for their approval. Also, these written words must be presented cogently to appeal to the students who will discuss the situation in the classroom.

The case is a multiplier, not an end in itself. If the case researcher had only a handful of students, and unlimited time, he could take his students with him into the field to study administrative situations directly. With hundreds of students this is manifestly impossible. The written case, therefore, is the first step in recreating for large numbers of students the experience of the case researcher in the field.

Scores of visiting professors year after year listened attentively to "the voice of experience" as he talked about the literary task of case writing. The voice was Malcolm P. McNair (Lincoln Filene Professor of Marketing, Emeritus), who had come to the Harvard Business School in 1920 and had taken an active part in every phase of the School's work. His recollection of the School's early days was always of interest and has served as an important background for many parts of this book. Of particular help to participants were his ideas on case writing. After these ideas have been noted, attention will be directed to the feedback which participants received from their coaches and colleagues as they put the data from the field situations into successive drafts of their own written cases.

The following excerpts are condensed from a videotape of Professor McNair's discussion during the Case Method Program in 1964.

> I think it's highly important that the case develop a great deal of *interest on the part of the student.* . . . This seems to me to be a very cardinal point in the case method. A year or so back I happened to run into a little story that was published in the Standard Oil Company's sheet sent around to stockholders. . . . This described an experiment that was made out here in the Newton High School by some of our people here at the Business School and some educators from other areas. . . . This was an endeavor to teach economics to teen-age people by the case method. . . . As the observer said: "This was the first time I've ever seen economics made fascinating to kids. I was amazed at the way they picked up economic concepts. The personalized approach got to them. It was like pursuing a detective story." I'll come back to that thought a little bit later in another connection, but this idea of the development of interest — getting students wound up in the situation and becoming concerned with the situation — I think is a highly important part of the case discussion.
>
> This is the sort of thing which I believe there's a literary phrase for — *the willing suspension and disbelief.* In other words, the willingness to take at face value the situation which the case presents, forgetting that this is artificial, so to speak, forgetting that this is a case, forgetting that this is a classroom, being willing to take the situation at face value and become the person concerned with it — that is the ideal that a case discussion ought to achieve.

Now, more particularly, with reference to the case itself (and you people are learning now about writing cases). Let's say some things that a case is not. The case is not just an example. It is not, on the other extreme, in my opinion simply a straight-out, photographic slice of life. It is perhaps somewhere between those things.

I think the instances of successful use of so-called "armchair" cases are pretty few and far between. . . . The "example," which goes back to the medieval *exemplum*, used by religious writers to express a moral precept, in my opinion is not what a case should be. . . . And a case is not, I think, a sheer guessing contest — a kind of a blind thing in which you repeatedly ask a student, "What is this?" or "What do you do in this situation?" "No, you're wrong — No, you're wrong." That is the sort of thing where you say: "Guess what I've got in my hand?" and after a number of people have guessed, you say: "No, that isn't right; this is what I've got in my hand." I remember a great many years ago when one of these "guessing" cases was presented by a businessman in the Business Policy course, and this type of situation came up. He presented the case of where he should locate a Mid-Western plant (he had a plant here in Boston), and most of the data which he presented pertained to two different localities. I've forgotten where they were, maybe South Bend and Dubuque, or something like that, and most of the data were about these two localities. He left the students to make a choice. They wrote their reports, and he came back about two weeks later and said: "No, you fellows were all wrong; the answer was Kewanee." There was nothing in his presentation that gave any clue. . . . That is not a successful kind of case.

And that is where the detective story analogy comes in. . . . Because you know that if you read a detective story about "who done it?" "who knocked off the old boy?" and so on, and you got all the way through and discovered that somebody from another state that you'd never heard of, that wasn't even in the story, did it, you'd say that that wasn't a very good detective story. You expect that there are going to be *clues* in there which will enable the sharp reader perhaps to figure out who it is that the author has designated as the villain in the picture. And if it's completely blind and there is no possibility of figuring out any clues, it's a lousy job. Well, in the same way, a case to which the answer is Kewanee and all the author has told you about is South Bend and Dubuque, well, it's a lousy case. So a case is not an armchair example, and it's not, I think, a completely photographic transcript — I'd like to come to that again in a moment — and it's not just a guessing contest.

I think that it is fair to say that *a case is really a distinct literary form.* It is obviously not a poem. It is not purely a narrative, but it has important elements of narration. It is not purely an exposition, though it has important elements of exposition. It is not purely argumentation, but it may have important elements of argumentation. It is not just fiction, but as I've just indicated, it may have an important analogy to a detective story.

I think that a case needs to be recognized as having a certain structure, an anatomy of its own, in fact, you might say, a series of structures. There is, for one thing, *a time structure* — a case takes place, a business situation takes place, in time. And there must be a fairly clear perception on the part of the student of what the time sequence was of the events taking place in this case. And you may organize a case around that. On the other hand you may find it necessary to break the time sequence at certain points in order to bring into the case some other considerations. But you need to be fairly clear as to what was the sequence of the events, what was the time structure of the case. From the standpoint of developing interest you may want to take as a point of initiation for a case a late event in the actual sequence, because that may serve to heighten the interest — how the problem first came to light, so to speak. Then you may manage through a flashback to show the early origins of the situation and build this up. But you always have to keep in mind the time structure of the case, and make sure that it is sufficiently clear to the reader.

Of course, as you already recognize, I am sure, the development of the time structure is a little complicated by the fact that cases typically need to be *written in the past tense.* A case that you have written currently as of this July is fine for discussion this fall, but even by next spring it has become dated; and it strikes the student as being a little bizarre if it is talking in the current tense about something that happened a year or two years or five years back. So, typically cases are cast in the past tense, and this gives you somewhat of a problem from a writing standpoint in keeping your tense sequence straight.

The case, I think, in addition to a time structure, needs to have a *narrative structure* — not only is there a time sequence of events, but the things that happened and the circumstances of their happening must be narrated in some kind of understandable pattern. Where did they begin, what led to what, and so on. In other words there is a flow of a story.

And clearly there has to be one or more *expository structures* in the case. There is the situation itself which the writer must make clear to the student, remembering that the student will not have the perception that the professor has. Thus, the case writer may have to spell the thing out somewhat more specifically for the student to grasp, than would be necessary for the group of business executives who participated in the situation. Clearly also there is an expository structure with reference to the company itself. Something must be said about the company context in which the action takes place. The student needs to have a good picture of the company. In addition, there is often an expository or descriptive job that needs to be done with reference to the industry . . . depending of course on the particular case. And then not infrequently there is a job that needs to be done on technological background. . . .

But I should come back again and stress perhaps most of all the importance of what I would call the *plot structure of a case.* The need for that was indicated earlier from the standpoint of developing interest and promoting the willing suspension of disbelief. For the case to be a really living thing and for the student to forget that it's artificial, there must be drama, there must be suspense. The skillful case writer will build this up. We for many years have used the shorthand expression that *a case must have an issue.* There must be a question of what should somebody do, what should somebody have done, who is to blame for the situation, what is the best decision to be made under the circumstances. A case isn't just a bland narrative where there's no question or issue. A case involves a problem of some kind, and the more you can build that up, the more you can develop the interest or the drama of that clash of ideas or maybe a clash of people, the better the chance of getting the kind of student commitment that makes for a good classroom discussion.

So, I think the case has a number of structures. It has a time structure, it has a narrative structure, it has several expositional structures, it has a plot structure. And frequently you have to interweave these in order to get the best results. You may start out with some incident which indicates the beginning of some kind of a situation, or perhaps the precipitating circumstance that brings the thing first to the attention of some executive. You may want to drop back from that in time to what has occurred earlier, or you may at that point want to go to the exposition of the type of company or industry, and so forth. These several structures usually need to be interwoven. This makes the development of a case, I think, a serious problem in literary composition. But, unfortunately, I don't think there are any textbooks written on this particular kind of composition.

Then there are what might be called literary problems and devices. I have mentioned a couple of times that the case is not just a photographic slice of life. A case represents *selection from a situation*; it represents selection by the case writer. Now this poses a problem; this makes it necessary, I think, for a case writer not to take just one man's view of a business situation, but to get as many different slants as he can from as many of the company executives as possible. But still the writer can't avoid coming back to the feeling: "I'm the case writer who is trying to make a selection of facts from this situation and to fit these into this rather complex type of literary composition which we call a case, and I've got to be very careful not to get my own personal feelings and judgments into the situation." Some people, in order to get around this particular difficulty, have used the device which you've probably seen in some cases of bringing the case writer himself into the case. Personally, I have never liked this, because I think this device runs counter to the student's feeling of reality about the case. As soon as the case writer himself comes into it, you immediately are reminded that this is an artificial piece of work; you lose that sense of the reality of the situation. Yet it is true that the case writer does play a very important part by his selection of the things that he includes in the case. Sometimes I think it is not stretching it too far to say that one of the questions here is, How much literary license do you have in writing a case? To what extent is it legitimate to heighten an interest by the process of selection? How far can you go? I think that sometimes it may be within the bounds of permissibility for a case writer to masquerade, perhaps as a consultant, or a friendly adviser, in the picture. That doesn't destroy the illusion and yet it does indicate that somebody is bringing in ideas that aren't necessarily those stated by the executives in the company.

You see, one of the problems here is that anything that is stated in a case, unless it is specifically attributed to some person, will be taken by the student at absolute face value. If you question the student on any unattributed statement, he says: "Well, it says so in the case." And I think he has a clear logic on his side. . . . But if the case says, "The *President thought* that," or "The *controller surmised* that," or something of that nature, whether it is a direct quotation or not, then the tendency of the student is to be on his guard and ask himself, "Well, do I accept this or not?" . . . If the statement is something which the instructor actually expects the student to question, I think it is important that the particular statement be attributed to someone in writing the case. That is, it should be indicated as an opinion, perhaps of an executive in the company or maybe of an adviser, the unofficial advice perhaps of a consultant, or something like that, in the nature of a sounding board. It is dangerous for a case writer just to lay flat out a statement which he really wants the student to question.

So there are some very ticklish problems in this business of making a selection from the material that you are dealing with and trying to put it in a form to arouse student interest and also to put it in a form that the student can really understand. This is why I say that a case is not just a photographic slice of life.

There is also this *problem of people* in the case. Years ago some of the teachers used to argue that we shouldn't let the personal side come into the case, just stick to the facts. But that was well before the reign of human relations began here at Harvard. A current illustration comes to my mind. . . . Mr. Sloan of General Motors, as you know, has recently written a book which was published serially in abridged form in *Fortune* and is now on the bookstands.[1] This is a distillation essentially of his experiences as the chief business executive of a great, perhaps the greatest, business organization. In the current issue of *Fortune*, Mr. Peter Drucker, a writer on business management subjects, takes Mr. Sloan very severely to task because in his narrative at all times he essentially said: "At this point, here was the problem we faced. It seemed to me that this was how it developed. There seemed to be these possibilities. This is what we did and why. This was the result." And then he moved on to the next one. At no point, says Mr. Drucker, did Mr. Sloan say anything about people and their feelings or how he persuaded his associates to do various things, not a word of that anywhere in his book; and at no point did he lay down any general rules of conduct by which he was guided in his management work. So Drucker was greatly disappointed with this book of Mr. Sloan's.

Well, I think our early cases at the School were somewhat like Mr. Sloan's book, at least in one respect. People and their feelings were not allowed to come into the cases. Of course, the thinking on that matter has changed a great deal in these days. But nevertheless while you are frequently inclined to bring people and personal feelings in, perhaps to use the device of dialogue for the purpose of heightening interest, remember that students are easily tempted to go overboard on this angle and concentrate their attention on people and personal aspects rather than on the analytical side of the case. So, while I think it is useful to bring personalities to some extent into a case if they are pertinent to

---

[1] Alfred P. Sloan, *My Years with General Motors* (New York: Doubleday and Co., Inc., 1964).

the problem, this practice courts the danger that the students will be looking for human relations problems in situations when other issues are more important.

Another of the literary problems in case writing is the *indication of an issue.* Quite a few of our cases — particularly our first-year cases — are likely to be fairly clear-cut in indicating an issue. For instance, I have a case here which is a very good case, a very useful case. The first paragraph reads as follows: "In April 1952, the executives of Western Auto Supply Company in Kansas City, Missouri, were concerned with the results of the second year's operations of the incentive compensation program for the company's retail store managers and assistant managers." The second paragraph goes on to develop the reasons why they were concerned: The amounts of executive bonuses had been very much greater in one year than they were in the following year, and because they had been lowered so much in the following year, the morale of the store managers had been badly injured. That is just a bald, straight-out statement. You know right away that the issue in this case is about an executive compensation plan which has failed to work out, that questions will be raised as to why the plan failed, whether it can be corrected, or whether it should be dropped, and so on. I think maybe in a good many first-year cases a bald statement of that kind may be useful. But there are many situations in which it is better to leave it to the student to analyze what he thinks the issue is. Thus the case will present the various precipitating incidents and pieces of evidence that indicate something isn't right, rather than coming out and baldly stating the issue. Here again, these are what I call literary problems in writing a good case. . . .

Another problem might be described as *bridging the gap.* There is still a great gap in comprehension between a student, even in a graduate school, and the business executive. The business executive has become accustomed to the use of a certain amount of shorthand, so to speak, in describing business situations — the report, memoranda and so on, that circulate in companies are frequently not usable directly in their original form in a case. Sometimes they are, but in a good many situations you've got to go a step beyond these raw data in your case presentation. Also, if you have a good deal of data in a case, remembering that the student has the job of preparing for two or three classes for the next day, it may be desirable to do some putting together of data for the student. If he is left the entire job of doing that, he may be so baffled that he does only a half-hearted job of preparation. Thus we sometimes find, after the initial experience of teaching a new case, that we need to take some steps here and there to bridge the gap, to put some of that information together in a more usable form, going beyond the form in which we got the material originally from the business executive.

There is always, of course, *the disguise problem.* Where you can get permission to use the actual company names and facts that is fine, but this is frequently not possible; and disguise problems are sometimes quite ticklish. If the company insists that you *disguise the industry,* it's probably no good; you might as well drop the case at that point. Sometimes, however, I think you can disguise the product. We have a quite successful case . . . in a large retail organization in which the product department, for purposes of disguise was shifted from one type of major appliance to another. It was exactly the same kind of problem, and I don't think anything was lost by making that particular shift in product. But if the executives want you to change the industry, it usually is no good. Commonly the disguise problem can be met by preserving the ratios and changing the dollar values, in other words, using some kind of constant throughout the case to alter the dollar data. Also the location can usually be changed. In your selection of a fictitious name avoid using one that might be the name of a real company. Sometimes you can leave a false trail; if you have, for instance, a case on Sears Roebuck, you might write it up to make it sound like Montgomery Ward, without in any sense really destroying the fact that you're dealing with a very large mail-order company.

There are various other minor writing problems such as use of headings, appendices, and so on. I don't think it's necessary to go into those; here you can use your own judgment.

Finally there is the matter of using questions at the end of the case. I have told you I personally don't like the device of bringing the case writer himself into the case because I don't like to destroy the illusion. In the same way, I don't like questions on cases. The minute you've got questions 1, 2, 3, 4, 5, 6, as you frequently see in the case books . . . you have destroyed the illusion. The student is made acutely aware at this point that the case is an artificial teaching device. But if the situation is presented to him in such a way that he is interested in developing ideas about what to do in this particular situation, and has become intimately concerned with his problems so that he is really putting himself into the shoes of a particular individual, that seems to me to lead to the most desirable atmosphere and most effective case discussion. So I sheer away from anything that pierces that essential veil of illusion.

I sympathize fully with the problem of the case writer trying to manage his material. As I have tried to indicate to you, case writing is an art, and a good case is a definite literary accomplishment.

Professor McNair's presentation explained how the case writer puts together a selection of primary data from the field: what people are doing in the situation; with whom they are doing it; and what it means to them, with minimum intrusion of the case writer's own evaluation. It might be called the classic issue case, presented around a more or less clearly identified issue requiring action.

Listening to Professor McNair was a little like watching a golf pro. It seemed so easy. In the Program everyone was busy trying it himself. And before long, drafts came in to the coaches, and their blue pencils created immediate cases in interpersonal relations. The written case, being tangible, provided a specific means for the author to get feedback not only from the coaches but from other participants who studied and discussed it. Professor McNair himself was one of the most effective coaches and superbly demonstrated both kindly encouragement to try the art of case writing and then critical blue-penciling of results. The critic does not create but can help the case writer see what has or has not been communicated, and he can point out the difference between facts and the author's opinion.

Maybe one of the reasons Harvard has developed as many case writers as it has is the fact that many of the cases have been written by eager and resilient young men working as assistants to men grown old and hardened in the belief that rewriting and editing and more rewriting have virtue in themselves — particularly if someone else is doing it. Now and then there appears a genius who puts a complex case situation into timeless prose at the first effort, but the work of the world is done by those who must try and try again. Care in case writing also is justified by the saving of time for literally thousands of students who will eventually study a good case. How far to press participants to rewrite their cases was a perennial question. Within the framework of the Program concept of the coach as host, and the eight-week maximum time, less revision was practiced than the coaches would have otherwise expected. They settled for encouragement rather than refinement of the participants' final efforts.

Although there are other case forms, the Program concentrated on the classic form because it emerges from and contributes to understanding the whole case complex of field research, writing, course design, and classroom discussion. Emphasis on this classic case form helped the participants see the interdependence of elements in the set.

As questions arose during the summer, each coach in his own way explained over and over again the basic ideas of this seemingly simple case form. A major difficulty in accepting it was the self-restraint imposed upon the case writer. Apparently this had to be repeated so often by John C. Baker, when he was a professor at the Harvard Business School and coaching this author 35 years ago, that his words and their refrain come to mind whenever a short explanation of how to write a case is needed:

> In the first paragraph you make a statement about the issue which must be decided in the company. In the second paragraph you give a little background about the company and its place in the industry. In the third paragraph, you tell how the situation requiring action developed. Then, as needed material is available, you give additional data about the way management handled earlier events leading up to the issue, and data from which alternative courses of action might be visualized. *You carefully leave out of the case your own opinions and evaluation.*

Each coach who worked with the Visiting Professors would give a different short description and acknowledge the wide range of useful variations in a consistent case method pattern.

Although a case is a distinct literary form, there is little room in it for "literary license." It is the connecting link between the professor's conceptual scheme, the classroom, and the concrete reality of the practicing administrator.

The Evolution of the "Classic" Case

At the Harvard Business School, the desire to involve students *in* a situation requiring analysis and action, rather than having them talk *about* someone else's theory and evaluation has long been a dominant concern of the Faculty. But the form of writing to facilitate this objective has not come without much trial and error. By such experimentation, classes were found to be more successful when the case stressed data and excluded evaluation by the researcher. The case typically stops at the point where action is necessary, so that in the classroom students may be projected into taking a position under conditions of uncertainty rather than being allowed to second-guess what someone else has already done. In arriving at this type of case many alternatives have been tried and some are still used on occasion.

The Faculty has tried to give students repeated practice in working with data, in discovering for themselves what needs to be done in a concrete situation, and in committing themselves to a course of action. This practice in relating data to a plan of action for a particular enterprise is one form of generalization. In the process of acquiring skill in relating data to action, the student begins to see, through a sequence of cases in a course, a second type of generalization: the flow of funds, the relation of volume to costs and prices, or similar subject-conceptual schemes. The time required to acquire such proficiency in analytical skill is quite different in scope from the time required to get intellectual comprehension of a subject. This difference in time was one of the major elements in the case method pattern which the Program tried to demonstrate.

A bit of history may help make the point. Professor McNair's views (mentioned earlier) came from more than 40 years of trial and error at the Harvard Business School where he played so active a part. The first case book in 1920/21 was a conversion of a book on marketing which the author, Dr. Copeland, made at the request of Dean Wallace B. Donham. Then followed about five years in which the Bureau of Business Research under Dr. Copeland's direction collected several hundred cases. The efforts of Dean Donham to get other professors to use these cases in their courses and the efforts of the Bureau to get other professors to give them specifications for the cases needed apparently led to mounting tension. By 1925 or 1926 it had become apparent that case research was an inseparable part of a professor's own course development, and at that time the centralized case program was decentralized. Reference to this transition is made by Dr. Melvin

T. Copeland in his book about the Harvard Business School.[2]

Excerpts from a memorandum of a discussion about case writing among staff members of the Bureau of Business Research show that many aspects of the classic case were already being recognized in 1925:

> Organization of Cases:
>
> (1) State issue first paragraph if possible.
> (2) Thread of issue runs through case.
> (3) Give pertinent facts germane to issue.
> (4) Also other information to show exact setting. Leave nothing to doubt.
> (5) Head notes do not take the place of well-written start to case.
> (6) General significance slips abolished. But use if section boss says to.
>   (a) Every problem not a case — must apply to other circumstances.
> (7) Some think too much about technique — not how to get across to businessman.
> (8) Should be transition to show how each succeeding batch of facts relates to issue.
> (9) Don't rely on businessman to bring up all significant facts.
> (10) Use active instead of subjunctive "would be."
>   (a) Makes for surer facts
>   (b) Cuts out arguments (particularly agents')

The conclusion of this 1925 conference memo epitomizes the commitment to understanding the business firm as the dominant reference:

> We have right to ask for facts and time of executives to help education of business. No businessman has leg to stand on if not agree. We are human sponges for business facts.

Following this outline came a wide range of comments from different staff members' experiences dealing with problems of meeting case "specifications," clarifying "head notes," and relations with the business executives. Many of these issues continued to concern Program participants 40 years later.

---

[2] *And Mark an Era* (Boston: Little, Brown & Company, 1958), pp. 227–238.

The stress between the "specifications" or hypothesis from a professor's course and what the researcher actually finds in the field situation provides the impulse from which one develops better understanding of administration. Subdividing responsibility for course concepts and for field research complicates this essential interaction. These two parts of the process were united 40 years ago at the Harvard Business School, but this bit of history is included because the idea of a centralized case research effort still comes up in American universities and frequently overseas. The experiment in centralization also helps to explain the problems which many professors new to case writing have in facing the stress between the ideas in their courses and the way the facts of an actual situation fit together. The act of case writinging also is the act of revising one's own concepts — a task all are a little reluctant to do.

The "head note" reference represents a different but continuing question. This has to do with the use of categories or concepts instead of the terms used by people in the actual administrative situation. The proposal was made by Professor Charles I. Gragg who in 1925 was Editor of the *Harvard Business Reports.* These reports were published by A. W. Shaw Company at first, and later by McGraw-Hill Book Company, in the same buckram binding used in the typical law office library of appellate court opinions. Consultants were brought in to help systematize the head notes and commentaries of these case reports. In the light of present comments about sticking to the facts of the case situation and minimizing the author's opinions, it may be of interest to note what a different approach was being tried in 1925.

A memorandum by Professor Gragg, dated July 13, 1925, started off with the paragraph:

> A Survey of several cases for tentative indexing has led to the following suggestions for the application of the research point of view to individual cases.

The memorandum goes on to illustrate how "more analytically chosen terms" would make the facts "more pointed and of greater general interest and significance." After illustrating two different ways of expressing facts in a case, he went on to say:

> The use of the more specific terms suggested would have made the case stronger . . . and also would have facilitated the task of indexing and digesting. . . . Include such specific words and phrases as would crystallize unmistakably the important topics involved. . . . The use of accurate terminology will require conscious endeavor to state the facts and to formulate the significance of those facts. . . . Terms should be chosen that analyze accurately the facts reported, thereby forcing on the reader a recognition of underlying principles which otherwise he would not have perceived because he would have thought that the facts of the particular situation had little or no significance for him. . . . The use of words and phrases which suggest conclusions of general importance and applicability should be an important phase in case preparation.

The changes since 1925 are dramatic for those who knew Professor Gragg in the post-World War II period.[3] In fact, in this later period Professor Gragg came to be identified as the ultimate in the "nondirective" method. However, having substituted for him in the classroom, the author believes that his classes had a sense of direction which any professor would envy. The students were given this sense of direction by his very consistent behavior in holding their attention to the facts in the case and at the same time allowing the students to explore what these facts meant in terms of action.

This may be enough to indicate that from years of trial and error the tradition was established of keeping the statement of the case facts as nearly as possible in the terms used by the business executives and others in the case situation. Thus, over the years, regardless of changes in their analytical propositions, succeeding generations of Faculty members have been able to utilize the facts of case situations as reported in the written case with increasing insight.

---

[3] Charles I. Gragg, "Teachers Must Also Learn," *Harvard Educational Review*, Vol. X, No. 1 (January 1940), pp. 30–47; Charles I. Gragg, "Because Wisdom Can't be Told," *Harvard Alumni Bulletin*, October 19, 1950 (reprinted in *The Case Method at the Harvard Business School*, by Malcolm P. McNair and Anita C. Hersum, New York: McGraw-Hill Book Co., Inc., 1954; and in *The Case Method of Teaching Human Relations and Administration*, edited by Kenneth R. Andrews, Cambridge, Mass.: Harvard University Press, 1953).

From these 1925 experiences in the field came the statement of the case facts for classroom discussion; and from the efforts of systematizing notes in the *Harvard Business Reports* came the research monograph. Out of the trials and errors of the succeeding 40 years came a clear differentiation between the classic classroom case and the case commentary. Both continued to depend upon intensive study of situations in the field. The case commentary evolved into notes on a series of cases about a particular subject, such as industrial marketing, and then evolved into the exposition of a subject, such as the economics of advertising, in which variations were documented by brief references to numerous case situations. It was recognized, as project research developed, that the resulting research report should describe the author's thinking about a particular concept or category of experience, with only enough documentation to permit understanding and independent confirmation or question.

Another branch of this evolution was the history of a particular industry or enterprise in which the historian's interpretation was interrelated with the time analysis. The finished documents were frequently called case histories, but their place in the educational scheme has been quite different from that of the written case described by Professor McNair.

More akin to the classic issue case were the so-called "postage stamp cases," which Dr. Copeland used in his first marketing casebook as a basis for starting discussions. Such cases were later refined into a systematic program by Professor Paul Pigors of the Massachusetts Institute of Technology. Professor Pigors frequently met with the Program to discuss what he called "the incident process." He was very effective in starting a class with a brief statement of an incident about which the participants would then ask for more facts. Thus, they recreated in the classroom an additional depth of involvement, roughly analagous to field inquiry by case researchers.

In addition to these forms of research reports, there also has emerged a great volume of business journalism in which are reported cases of success in use of forklift trucks, computers, air cargo, time study, work measurement, private brand merchandising, and so on.

As all these other types of cases continue to be written, there is need to help the reader when he picks up a case of the type that Professor McNair described. In an effort to keep the reader from looking in such cases for the author's opinion about what should or should not be done, or for an example of good or bad practice, Harvard cases have been documented to warn the reader that:

> Cases are designed for purposes of classroom discussion and not to illustrate either effective or ineffective administrative practice.

Distinction Between Case and Article

Part of the difficulty in case writing has always been the need for the literary skill which Professor McNair described in his discussion of the Case Method Program in 1964. This skill, like so many others, increases with practice. But hundreds of tremendously useful cases have been written by men who have made no pretense of literary expertise. In the Program, it became apparent that apart from lack of skill a much more basic difficulty was the failure to distinguish this particular literary form, the classic case, from the forms of scholarly writing to which professors were accustomed, such as theses or journal articles.

Many of the Visiting Professors seemed to be hindered by the fact that they were accustomed to developing and presenting their own thinking and opinions in lectures and journals. Even when they began to write a case, they were preoccupied with what had been in their minds before they went into the field. They wanted to write the case in such a way as to communicate some of their course ideas. Frequently they wanted to bypass identification of the issue and settle for a document which would help them and their students *survey* in concrete terms the topics covered in their courses. In their field research, they were not placing a high value on observing or interviewing the administrator, or even on testing their own ideas. They were placing a high value on the exposition of their course concepts, and still thinking in terms of a lecture or a section of a text. For them it was very difficult to discard their customary way of organizing ideas in order to state the facts of a given situation.

In contrast, the case writer was asked to submerge all these interests in favor of the *data* which students can use to expose their own thinking and test their own opinions in class. In the case method, the place for the exposition and discussion of course concepts is in the discussion of the case, not in the written case itself. In writing the case, the author must discriminate constantly between those aspects of the field situation which are to go into the case and those which are to be recreated in classroom discussion or otherwise.

In brief, the case method was and is as much concerned with *what is omitted from the written case* as with what is included. This distinction was repeatedly discussed between coaches and participants in the Program, and much of the coaches' work was to persuade the case writers to leave out of their cases what sometimes seemed most significant to them. To insist that the writers' opinions be omitted is not to deny their importance, but to stress that they come into the case method pattern in a different way.

For a case writer to make separate note of his own ideas about a situation while writing the case itself is a convenient way to sort out what actually does go into the case and what does not. This results in the "teaching note,' the last of the four documents created in the process of case writing: (1) The researcher's journal in which are recorded the raw data of observation and interviewing; (2) The worksheets and outlines made in the process of analyzing and discovering what the data mean; (3) The case; and (4) The teaching note. The distinctions manifest in these documents suggest the interrelationships of all the elements in the set of activities called the case method.

Teaching Note

The repeated denial of a place for the researchers' opinions in the written case does not mean that there is no place for them in the case method pattern. On the contrary, it is for the clarification of the professor's judgment that he goes into the field. Therefore, many of the Visiting Professors found it helpful to start a teaching note for the case along with writing the case itself. Thus, they had a place to note their own ideas about the situation when these were edited out of the case draft. It is imperative for the researcher to record his own analysis in some separate document. Getting participants to write both the data and their analysis of them was a major diplomatic effort of the coaches.

When the case researcher had finished analyzing the situation and had satisfied himself with the implications of various choices of action, he clearly felt an increase in his understanding of administration. One concept of education would prefer that the case researcher then publish his analysis with supporting data. The educational philosophy of the case complex, however, considers that the creative and developmental exercise is in the analysis of the case data and in making an appropriate choice of action. The objective, therefore, is to multiply the experience of the researcher by putting only the facts of the case before students, and to give the students an opportunity for analysis similar to that which the case researcher himself had in the field.

Recreating for students the researcher's surprise of discovery in the field situation was one of the major contributions the researcher sought to provide in selecting and organizing data for the case. And the invention of the classic "issue" case has gone far to feed into the classroom the excitement of actual field research. As Professor McNair pointed out, the "issue" or point at which the responsible man in the case can make a difference by his discretion is the crux of the case document, whether it be explicitly stated or only implicit in the data.

If the issue or action point was not discovered in the field situation, the effort to insert the issue into a case represented one of the most dubious forms of "literary license," because the situation at the point when and where a decision should be made was what the researcher was trying to find. Paradoxically, however, the executive might not be aware of the issue at the time, any more than the driver who unwittingly passes a critical fork in a road. When the researcher has caught the administrator in the midst of an actual decision-making situation and has been able to put the relevant facts into a case, then the case becomes a "classic." Such cases have been used repeatedly with increasing insight by innumerable professors as they integrate the situation into the body of knowledge in their courses and discipline.

By being faithful to the facts in the situation and by excluding his own thinking and opinions, the case writer makes a more lasting contribution than he often realizes — to the executive in the situation as well. Seeing the written statement of the need for action which he faces helps the administrator get a balanced view of his own situation. Also, the researcher by reporting the set of facts in which he found his present understanding of a subject to be relevant or in need of change, stimulates further course development. Each student is led to expose his own thinking and feeling about a single set of facts; and thus he and his fellow students acquire greater awareness and better understanding of personal differences in administration.

The information in a case rarely is of lasting intrinsic value. Students do not study the case to *learn* the information in it, although they acquire as a byproduct a useful perspective of the business world. In writing a case, therefore, the professor seeks to design a means to the end for the students' learning. The objective is to help students gain skill in seeing a situation as a whole and in seeing how their special fields of interest can make a difference as elements in the situation. Nevertheless, it is of paramount importance to get the facts straight. The facts in the written case then provide a base from which classroom discussion can develop. These facts also provide a sampling of the context in which students will be living and working, and from which they must learn the possibilities of creative action.

Emphasis on the difference between data and thoughts about the data surely does not minimize the importance of the latter. Thoughts about the data constitute the reason for doing the research. To learn anything from field research, the researcher must be prepared to consider the need to change his course objectives and concepts if they do not allow him to deal adequately with the situation he finds in the field. To allow for modification of the relevant concepts, the facts in the case must be kept straight. A log of the researcher's thinking and feeling about the situation may be even more important for his own development than the case itself.

Cases frequently are criticized for being too dull. Not many achieve the level Professor McNair described. In the summer programs, new case writers pled for introducing a little dialogue into the case even though it had never been heard, or for reporting an interview never held, or for picturing a product never seen. But in the interest of professional development, the Program held to its goal of reflecting as closely as possible the facts in the field situation. A professor can always create any kind of an exercise for the purposes of his class, and of course the researcher can let himself go in his own teaching notes to record all kinds of ideas for new projects or possibilities to explore, but the drama of the situation is in the issue and the confrontation of students coming at the issue from different points of view, reacting to different aspects of the case.

Structural Aspects: Authorship, Ownership, Release

What does case authorship mean? At the time when the first Visiting Professors Case Method Program was being designed, the Committee on Case Development Policy of the Harvard Faculty, under the leadership of Harvard Business School Professor Edmund P. Learned, was nearing the end of a three-year-long series of conferences attended by a broad cross-section of Faculty of all ranks. The deliberations of this committee were frequently brought before and discussed by the Program. In the final report which was adopted by the Faculty, this committee recommended, among other things, that

> . . . the practice of writing cases anonymously be modified to permit appropriate acknowledgment of authorship on individual cases at the discretion of the member of the senior Faculty in charge of writing the specific case.
>
> The conditions which originally made anonymity of authorship desirable now seem to have become less important than the advantages of recognizing case authorship. Recognition should encourage case production and improvement of quality. The adoption of this practice at the Harvard Business School also promises a helpful precedent for case writing in other schools.
>
> Cases are prepared under widely varying conditions. There-

fore, in noting authorship, three elements need to be taken into account in each case.

(a) *The business firm* that provides the case material. Frequently the company prefers to remain anonymous.
(b) *The professor* whose objectives and course-development concepts provide the basis for selection and arrangement of case data.
(c) *The case writer* whose creative efforts produce the case in a form meeting both the approval of the company and the pedagogical requirements of the professor.

Experience will indicate appropriate types of acknowledgment. The following statements are offered as suggestions. Members of the Faculty who feel that other types of authorship acknowledgment better fit their circumstances are encouraged to bring their statements to the Director of Case Development for his concurrence, e.g.:

> This case was made possible by the cooperation of a business firm which remains anonymous. It was prepared by Mr. John Doe, Research Assistant, under the direction of Professor James Roe as the basis for class discussion rather than to illustrate either effective or ineffective handling of an administrative situation.
> Copyright © 1956 by the President and Fellows of Harvard College

The Visiting Professors generally followed some such form of footnote acknowledgment of authorship. The copyright practice varied from university to university, some of the state schools being required by law to put all research efforts in the public domain.

Practice in the Program was to have the professor take his new case to a responsible officer in the company for his "release." The constructive relationship of confidence which normally characterizes case research was symbolized by this submission of the finished case to the informants to see whether or not the essential relationships were in focus and the nonessential identifying background blurred enough to assure privacy. Hesitation of the executive to release the case may be a clue that the author has exposed more of the company than is necessary for the analysis and decision to be considered in this particular case-class discussion. Disguising the identity of the company is made more difficult when often distracting and simply interesting details are included, especially when they may be only remotely related to the immediate issue. The disguise and discarding of unnecessary detail in a sense is like the sheet drawn over the rest of the body in the operating room while surgeons and interns are examining the area of operation.

The form for case release developed for use in the Program was a letter (in duplicate) sent by the professor to the company with the finished case. Some such paragraph as follows was often used:

> I appreciate your cooperation in making your experience available this summer, from which we have prepared the attached case entitled "——————————————." I trust that it captures the important elements in the situation and at the same time avoids divulging any information which should be kept in confidence. We believe it will be very useful in teaching. When you have noted any corrections, will you please sign the carbon of this letter to show you approve publication of the case and its distribution for educational purposes by the Intercollegiate Case Clearinghouse.

The duplicate, when the case was approved, was endorsed and returned so that the case could be contributed by its author to the Intercollegiate Case Clearing House for publication and widespread distribution, along with a supplemental note for other teachers. Having excluded his opinion from the case, the author could now let other professors who might use the case have some clues as to the point of view from which the case was written. This was not to give "the answer" to the case, but to help the professor who is building a course of his own know the possibilities of the new case.

Exhibit III-1 lists the information which is needed by the Intercollegiate Case Clearing House in order to list the case and make it available for widespread use (discussed further in Chapter VII). Preparing the case information sheet so often helped visitors to clarify the focus in a case that in later years coaches asked for a preliminary information sheet *before* the case draft was started. Filling it out brought to attention many structural issues of the sort described by Professor McNair.

The availability of cases through ICH must not mislead a pro-

EXHIBIT III-1

INFORMATION TO ACCOMPANY CASES

(Please make entries specific and complete enough to help instructors order cases from your listing.)

Section of Bibliography where case is to be listed (check one):

Controls.......... Finance.......... General.......... Human Aspects.......... Marketing.......... Production..........

IDENTIFICATION

Case title.. No. pages....................

Author..

Supervisor (if applicable)...

Institution..

Address..

Source of data (e.g. field, library, etc.)............ Languages of case: English............ other............

Supplementary material (if any) needed by students...

Before you sign, please note:

INFORMATION ON THIS SHEET GOES INTO THE BIBLIOGRAPHY. IT MUST NOT CONTAIN ANY CONFIDENTIAL MATERIAL. Be Especially Careful in Abstract Setting, Subject and Commentary

CERTIFICATE

In submitting this case to the Intercollegiate Case Clearing House (Policies on reverse of this sheet) I CERTIFY THAT AUTHORIZATION HAS BEEN RECEIVED FROM SOURCE FOR USE AND WIDESPREAD DISTRIBUTION.

Signature..

Title..

Institution...

ABSTRACT (What case is about and how it was used in class.)

....

....

....

....

....

....

COURSE FOR WHICH WRITTEN..

Section of course in which used..

Other subjects by which to index the case so others may see potential uses:

MAJOR SUBJECTS INVOLVED	Other Subjects
1.	....
2.	....
3.	....

SETTING OF CASE

a. Geographic.................... b. Industry.................... c. Size....................

d. Department (e.g. Sales, Engineering, etc.).................... e. Year of case event....................

f. Positions involved.................... g. Locale (e.g. factory, office, etc.)....................

Send information sheet and 10 copies or more of case, which can be distributed as samples	to	Grace V. Lindfors Intercollegiate Case Clearing House Soldiers Field Post Office Boston, Massachusetts 02163

fessor into thinking that he can take any three dozen cases and string them along into an effective course. Chapter IV will describe the activities of the Program in introducing participants to case-course development. Participants were always eager to obtain the further recommendations of the Case Development Policy Committee who dealt with the use of cases in casebooks. This committee's report noted that *all cases belonged to the university,* but in accordance with long-standing practice:

> The committee recommends that any member of the Faculty who has been active in case-course development work and desires to publish or revise a commercial case book request permission from the Dean. . . .
>
> . . . While each decision must be made on its merits, it is suggested that the Dean take account of the following criteria in reaching his decision:
>
> (a) The educational need for the proposed case book.
> (b) The potential value of the experience to the Faculty member concerned.
> (c) The equity of the proposal. In judging equity, it is suggested the Dean take account of prevailing Faculty sentiment that those Faculty members who are most active in developing and producing case and course materials should be considered as logical parties to whom permission be granted to publish case books or subsequent revisions.
> (d) The effect of the proposed case book on the reputation of the School and on the case method of instruction.
> (e) The effect of the proposed case book on the utility of the cases concerned in the Harvard Business School classrooms in the various programs.
>
> The committee considered the possibility that faculty members of other schools might ask for permission to use Harvard cases in a printed case book. While questioning, in general, the advisability of granting such requests, the committee recognizes the wide variety of possible circumstances and would leave discretion to the Dean in making such exceptions as were compatible with the criteria listed above. . . .
>
> Recognizing that the preparation of a published case book is typically done over and above a Faculty member's normal School assignment, the committee recommends that the Faculty reaffirm the policy that royalties accrue to the Faculty member(s) producing published case books, except in unusual situations where other

arrangements may be deemed appropriate by the Dean and the author(s).

Since the written case is a tangible manifestation of the whole case complex, writing a case brings to a head a wide range of issues from the essential nature of administration, language, and learning, to the technical details of publication and distribution. The written case draws on science but also is a work of art. The literary achievement must satisfy three different critics:

(1) The executive, who must confirm the basic reality of the report and be satisfied that confidential data have been properly used;
(2) The students, who must accept the situation as a challenge to their developing management skill. (See Chapter V.)
(3) The case researcher-professor himself, who must find a way to live with his concepts and this sampling of actual things, people, and events. The case must find a place in the professor's course. (See Chapter IV for case-course development.)

Chapter IV

Developing a Course with Cases and Concepts

PROGRAM PARTICIPANTS soon became involved with company executives and developed enthusiasm about discovery of the new cases. In addition, participants became increasingly aware of the fact that in separating the data for the case from their feelings and beliefs about the data, they were learning about themselves as well as the case writing experience.

Fairly early in the Program, the participants also recognized that perhaps some element of this involvement and self-awareness might be recreated in the classroom when discussing a case. Then there came the honest perplexity about how such experience in the field and in the classroom might be used to cover the substance of the courses for which they were responsible. (The classroom experiments will be discussed in Chapter V.)

This chapter has two sections. One analyzes the conflict participants seemed to feel between the contribution cases promised toward broad educational objectives and the coverage of subject matter for which they felt responsibility. The situation was complicated by the investment of time and anticipated effort to discover appropriate cases directly from the field or from the work of others. Strategies for resolving this dilemma by making the case course an instrument of active learning for instructors as well as students merit further testing.

The second section describes a selection from experiments to give participants experience in case-course development. This selection reports considerable classroom discussion to bring ideas of course development down to cases and to serve as a transition to Chapter V on the use of cases in the classroom.

SECTION ONE

Selection and Sequencing of Cases

Between the completion of a case and its use comes a critical step: the selection and sequencing of cases into a course. To the observer of a case discussion, the role of the faculty may seem very passive, but the casual observer has no way of knowing the amount of care which has gone into the selection and arrangement of cases for discussion. Here is the point at which the professor's knowledge of the subject and the objectives of the total curriculum come into play. Here is where the professor takes the center of the stage.

In the field the professor-researcher follows what is important to the executive. In the written case he cuts out all his own opinions. In the classroom he follows what is important to the student. By the use of cases professors gain new freedom to organize their own knowledge and interests in relation to the administrator's and the student's. However, this relationship is far from obvious, as repeated trials and errors in the Program demonstrated.

As has been shown previously, only a small part of the total field experience goes into the final written case. This field experience is used also in determining where the case might fit in the course. The field work also tests what background the student must have to enable him to feel the sense of discovery which made working with this data important to the researcher. Such considerations frequently result in a new conception of a professor's course.

In the process of field work and case writing, visiting professors began to see how relevant the idea of a case was to a business enterprise evolving in its environment through a sequence of decisions. They began to see the forces determining the direction of the enterprise and at the same time could appreciate the tension felt by the administrator at the point where he saw alternatives and the need to exercise discretion, to make a choice, a decision, to create a new course of action.

It seemed much more difficult for professors to think of making their own subject or course more relevant at such decision points. They often did not see that what they were teaching also was an evolving system of determinants with points at which an administrator might exercise discretion and redesign or modify the financial structure, social structure, wage payment system, marketing system, accounting system, or whatever aspect of business and its environment might be under study. The written case could be seen in the context of the company more easily than in the context of the course structure. It was only as participants in the Program began to face this crucial task of case-course development that they saw the connection between some of the points which the coaches were raising about field work, case writing, and the discussion of cases in the classroom. Using a single case does not require much change by the professor. Developing a case-course *does.* Implied here is the truism that what the instructor does in class depends upon his course objectives and, conversely, the design of his course depends upon what he plans to do in class. Similarly, much of what appears in this chapter also belongs in the discussion of the classroom, and what appears in the next chapter on the classroom is relevant to understanding course design.

The discovery of a case in the field and the increased awareness from classroom discussion made participants favorably disposed to learn more about developing a course by using cases. However, participants soon saw that no matter how much a person knows about his subject (marketing, production, finance, accounting, or something else), if he is to use cases in his course he must invest a substantial amount of additional time in order to master each new case. Participants saw how long it took to discover just one case in the field and how much more time it took to write it. They also saw how long it took to analyze each case to be discussed in class. Assuming the need to analyze a case for two out of three class periods each term, they foresaw how long it would take to master cases for each term for each of four or five courses which they had to cover. The time demand loomed ahead as a discouraging if not insurmountable barrier.

How then can a professor avoid this predicament and make his course an instrument of active learning for himself as well as for his students? The students will be under the same pressures; they will feel compelled to invest time in building up a backlog or reserve of cases by careful analysis. Practicing administrators have similar contraints; they need to learn more and more about different kinds of situations and, relative to their needs, have a limited backlog of case experience. Business administrators, students, and professors all share the need to understand the process of getting to where the action is — where in a system of determinants there is an opportunity for discretion. Practical precedents are not reliable; good intentions are not enough; abstract generalizations are not relevant.

It was a novel idea for most participants to think of organizing the concepts of their courses for the purpose of helping administrators and potential administrators to discover points for action in real life situations. It was even more thought-provoking to consider that this very capacity to convert abstract ideas into patterns to guide inquiry might be the most important thing students could learn while taking a course.

Most participants carried responsibility for so many courses that they felt it was unrealistic to think of creating a course of their own. Exhibit IV-1 shows the variety of courses for which participants in the Program were responsible. Most professors *adopt* textbooks or perhaps case books as the backbone of the three to six courses for which they are responsible. Book salesmen, therefore, provide a major channel of communication within the academic world. Professors find that there are practical pressures favoring the selection of a popular book of cases or texts. One publisher said: "You know, you made it awfully hard for us textbook publishers. Men came back from your Program and threw out the textbooks and began to use cases like mad. But, before long, they began to bring the texts back in. We frankly aren't looking for case books. We want authors who have developed a good text and introduce a few cases along with it for illustration. We want a textbook which will sell twenty to forty thousand copies." An alumnus of the Program, who enjoyed teaching and had moved to a university where this skill was valued, spoke frankly of his attention to the representatives of commercial publishers. He learned from them the way different

EXHIBIT IV-1

Illustrative Course Responsibilities of Participants in Visiting Professors Case Method Program

Professor A
- Management
- Advanced Management
- Administrative Problems

Professor B
- General Business
- Industrial Organization and Administration
- Personnel Management
- Business Policy

Professor C
- Principles of Management
- Wage and Salary Administration
- Human Relations in Industry
- Industrial Purchasing
- Production

Professor D
- Industrial Relations
- Collective Bargaining
- Business Policy and Organization

Professor E
- Wage and Salary Administration
- Production Planning
- Quality Control
- Operations Research
- Industrial Management

Professor F
- Business Finance
- Investments
- Accounting
- Money and Banking

Professor G
- Principles of Economics
- Economic Theory
- Current Economic Problems
- Financial Management
- Insurance

Professor G (alternate years)
- Business Policy
- Personnel Management
- Property and Casualty Insurance
- Corporation Finance

Professor H
- Accounting
- Accounting Problems in Management
- Finance
- Accounting Controls

Professor I
- Elementary Accounting
- Intermediate Accounting
- Cost Accounting
- Managerial Accounting
- Institutional and Governmental Accounting
- Financial Statement Analysis

Professor J
- Introduction to Marketing
- Industrial Marketing
- Marketing Theory
- Marketing Research

Professor K
- Principles of Marketing
- Wholesaling
- Sales Management
- Marketing Management
- Corporate Finance
- Business Policy

Professor L
- Principles of Marketing
- Sales Management
- Retail Business
- Credits and Collections
- Marketing Research
- Advertising

Professor M
- Marketing
- Transportation
- International Business

books had gone with different groups of students in other universities. As a result of this investigation, he and his dean selected a book which would be well accepted by their particular student body and an author who would be a credit to the department. The publisher mentioned above also said: "A good textbook starts with an awareness of concrete experiences the student has had and relates knowledge step by step. It leaves room and incentive for the student to become involved with the ideas in relato what he already knows." He would agree that a good textbook also would leave room for the teacher himself to become involved. Under such teaching loads and prevailing expectations, the extent to which a professor could become creatively involved in development of several courses at once had been so slight that for most participants the exercise of case-course development was a surprising innovation.

The program partially recognized that the task of case-course development defied its format of direct experience which had proved to be so effective in field work, case writing, and classroom case discussion. The *time* it takes to simulate case-course development was always more than the two months available in the Program. All attempts to sample the case method ran into questions of time. Indeed, formal education might be thought of in terms of time manipulation to telescope into a few months or years what a person might not encounter in the way of experience and ideas short of a lifetime. In a business policy case an instructor might cover in three hours of study and class discussion the climax of 50 years of corporate evolution, with decisions on new directions dramatically affecting the next 50 years. Or, in a human relations case, a professor might spend three hours of study and class discussion analyzing the implications of a five-minute telephone conversation as an index of the interpersonal relationships in an organization. There is then this inherent discrepancy in time scale, between the educational world and the business world, which the use of cases seeks to relate.

This time discrepancy was a very current problem in the operation of the case method program. Professors, as participants in an educational program, had the common mixed emotions of wanting to shorten the Program even while recognizing that there was still much more to master. The greatest time discrepancy came in attempting direct experience in case-course development. Within the eight weeks, visiting professors had a reasonable "one-to-one" "real time" transfer of the hours it takes to get data and to write a case. They had leadership of one or more class periods, again in about a one-to-one "real time" transfer. But even in eight weeks the Program could not make a "real time" simulation of case-course development over a nine-month academic year.

Frequently, the most important characteristics of case-course development turned out to be those intangible, personal, subjective elements of learning which required the longest periods of gestation or practice.

Since the Program had only eight weeks altogether for the whole range of activities and since even a one-term course develops over a period much longer than eight weeks, the exposure of visiting professors to the way in which a course develops through the use of cases always left much to be desired. In this course development activity, the Program had to fall back much more upon talking about the activity than in giving each man an opportunity to design, try out, and get feedback from his own case-course.

A coach can take a professor into the field and in a few hours or days they can actually discover a case together. It may take weeks to write the case. But development of a course requires months and years: time to master many cases, to select from this repertoire or reserve, and to sequence the cases, which with appropriate concepts make up the course.

A professor's resistance to converting his course to the use of cases is understandable and proper because he has responsibility for his students and needs to know what he will be subjecting them to. In encouraging case research we can insist that the way to begin is to begin. However, before undertaking to teach a course to students by use of cases, an instructor must be more clear about the contribution of each case to his course objectives, and the place of each case in the total sequence.

Basic preparation for case-course development in the Program was for the professor to get out in the field and to see for himself what the connection was between his course concepts in the abstract and the concrete situation faced by someone who might use these concepts. But it takes more than one such case experience to make a course, and there is risk in underestimating the time and effort necessary to get such experiences in adequate number to test and revise the conceptual scheme.

It was clear to the faculty even in the preliminary discussions with The Ford Foundation that two months was too short a time to provide participants with direct experience in organizing their courses around a sequence of cases. No matter how much the Program was appreciated each year by the participants, each year the coaches were least satisfied with this element of the Program.

Exposing visiting professors to course development also put strain on the basic idea of "host." It was difficult to avoid the role of teachers "giving" a course or of having visitors in the position of "taking" a course from the coaches. *But the visitors usually were not acquainted with enough cases to "give" a case-course themselves until the Program was over.* Therefore, efforts to simulate case-course development left a zig-zag path (see Exhibit IV-2) as the Program each year tried to emphasize what the previous participants said they failed to get in the preceding one. Within the limit of eight weeks adequate emphasis never could be given to all aspects. Fortunately the individual professor does not have quite this limit in his own course.

The systematic schedule of observations in 1955 led participants to recommend more latitude for picking and choosing cases to be observed. The 1956 participants, however, felt that they did not understand the way case discussions were developed into a course; so in 1957 participants were enrolled in an actual Middle Management course. But at the end of the summer of 1957 the visiting professors felt that enrollment in an actual course was not the most effective way of giving them direct experience in the activity of case-course development. So in 1958, more emphasis was put on "talking *about* case-course development." That was not enough, either. In 1959 another experiment was tried to let participants "experience" at least a brief course. Later in this chapter it will be described at some length. But in the following year, 1960, because the 1959 par-

EXHIBIT IV-2
VARIATIONS IN EFFORTS TO SIMULATE CASE-COURSE DEVELOPMENT

Emphasis on "Taking" a Course		Emphasis on "Giving" a Course	
1955	Followed one course in the Middle Management Program		
		1956	Compared observations in several Middle Management Program courses
1957	Participated in Policy course with Middle Management Program		
		1958	Compared observations in several Middle Management Program courses
1959	Participated in special course "Management in a Social System"		
		1960	Groups of participants developed five-session courses
1961	Became students in special course in Marketing		
		1962	Selected cases for participant's own course
1963	Participated in series of 14 cases, led by participants in teams		
		1964	Each participant taught a pair of cases
1965	Participated in one-week course in Marketing		

ticipants had questioned time taken by "the course," more brief "courselets" were tried. Then the pendulum swung back in 1961 to a "real" course and alternately in the next year away from a course. Such alternation continued, as shown in Exhibit IV-2, in one form or another throughout the remaining years.

In spite of limitations in the simulation of case-course development and the seemingly insuperable task of mastering an adequate reserve of cases for a course, participants persevered. Men were intrigued by the possibility that the use of cases might somehow permit them to progress toward realizing all the complex and compound objectives of education for business administration. Each summer at least ten professors from Harvard who were experienced in case-course development would bring in course outlines and discuss with participants their conceptual schemes. They would explain their experiments with selection and sequencing of cases and how readings, lectures, and technical notes were interspersed. The questions of case-course design are ones on which some professors spend a lifetime and as emeriti[1] are still emotionally involved. It became apparent in discussion of case-course design that questions involved clarification of purpose. Repeated exposure to Faculty members concerned with different aspects of administration left participants with the common realization that all of them shared a sense of responsibility for meeting the needs of a concrete enterprise as well as for covering the concepts of their particular course. Participants were also impressed in these seminars on case-course design that Faculty members shared responsibility for helping students develop as individuals in the process of working on the cases in the course.

Before exploring some of the ways in which the use of cases might develop a course, a brief review of objectives is indicated. The critical word in all these statements of purpose has turned out to be *AND*. It is the administrators' need for knowledge *and* skill in his concrete enterprise *and* in the world of science; for

[1] See "Development of Outline for a Course Taught by the Case Method," by Neil H. Borden, in *The Case Method at the Harvard Business School,* by Malcolm P. McNair and Anita C. Hersum (New York: McGraw-Hill Book Company, Inc., 1954), pp. 164ff.

discovery *and* for testing; *and* for maturity which makes the investment in cases worth the effort. In order to illustrate the depths and range of objectives, four aspects are considered: (1) Professional Orientation, (2) Generalization, (3) Skill, and (4) Maturity. But, if one has a different professional objective, these issues of case-course construction may seem very remote.

The importance of clarifying a participant's purpose was particularly stressed by Professor Ross Walker, who in the years before his retirement became almost a legendary part of the Advanced Management Program at the Harvard Business School. His ostensible course assignment was in Costs and Finance, but his interest was in the whole man. In his meeting with the visiting professors in 1958, Professor Walker was politely critical of the Program for possibly making the visiting professors concerned with courses as ends in themselves instead of with the critical need in the world for training executives who would take responsibility for making decisions.

While modestly minimizing his own part in the development of the case method, Professor Walker described the vision which dominated the early years of the Harvard Business School as not just concern for a method but for a *new kind of education.* He therefore urged the visiting professors: "Keep the beacon light of purpose before you at all times."

Orientation: Administrator and Expert

It was clear that for Professor Walker the purpose in his course was not just to communicate the generalizations at which he had arrived for budgeting or for financial management. He felt a genuine responsibility for helping students clarify their orientation as business administrators and for helping them develop skill and power in policy decision making. If a Program participant did not share his purposes, but intended primarily to communicate knowledge already established, quite different materials, methods, and course organization would follow.

In a professional school the students are in process of becoming administrators. By contrast, in the typical graduate school, where the students are in process of becoming scholars and teachers, there is little stress between the objectives and professional attributes of students and professors. However, there is a marked difference in career expectations between students and faculty in professional schools of business administration and even in undergraduate programs. In such schools, the student sees more or less clearly, and is more or less encouraged by faculty awareness, that his assimilation of the course materials needs to be different from that of his professors.

Rarely is the professor in a school of business administration concerned with the total life of the practicing administrator. The professor isolates some part of this totality. He often finds it easier to identify with the expert than with the administrator. He abstracts out marketing, production, finance, or some other sub-system of the enterprise or its environment or their interaction, for the subject of his concentrated attention. Or he may continue his career in some other academic discipline such as sociology, psychology, or mathematics, leaving their utilization in business to the students.

Thus, although there is a large measure of overlap between the professional interests of the administrator and the professor of administration, there is enough difference on important dimensions to merit clarification when undertaking the development of a course for students of administration.

In selecting participants, the Program favored a nominee who was at least willing to explore the possibility of designing courses to meet the needs of his students as potential administrators.

Chapter II described how the professor and the administrator found a community of interest at the point where a decision can make a difference. The case event is common both to the concrete system for which the administrator is responsible and to the abstract system of knowledge for which the professor is responsible. Thus, the case provides a meeting point for men with different points of view, identity, or professional orientation.

The scholar, who prides himself on knowing what he knows and what he does not know and on his adherence to the discipline of his profession, has a different orientation from the business administrator. The business administrator likewise benefits by knowing what he knows and what he does not know, but comes to accept responsibility for directing his enterprise even in the face of uncertainty. Whether or not modern science has the answer to a problem, the administrator must still take responsibility for keeping the enterprise alive by taking advantage of each opportunity.

A professional business school cannot simply concentrate on the isolation of some phase of knowledge which a professor would like to develop. This type of school must also prepare students for undertaking responsibility in complex enterprises. The use of cases provides a way of making progress toward both objectives. The changes in sequence and the balancing of lectures, notes, cases, and other materials can be better understood by considering some of the experiences of the visiting professors in the conduct of case class discussions. The use of cases makes the course a stimulating sequence of meeting points between students moving toward professional orientation as responsible, skilled administrators and a professor clarifying his professional orientation as a scholar in a particular field which is relevant to the administrator.

Perhaps it is only fair to warn a professor that his set of lecture notes or favorite text sequences may have to be drastically rearranged. When he collects a set of cases and starts to use them as building blocks, he frequently finds that he also needs to reconsider his generalizations and a new design of course concepts for establishing policy direction and decision in a specific enterprise.

Generalization: Direction of Inquiry and Direction of Enterprise

Of particular importance in the development of a course using cases is the need to distinguish between the types of generalization to be drawn. The administrator in a specific case has only a passing interest in the professor's subject, but he does have a continuing interest in the direction which his policy decision is going to take the enterprise as a whole. The professor has only a passing interest in the administrator's particular enterprise, but he does have a continuing interest in what the alternatives in this decision mean to the evolution of knowledge in his subject.

To help the administrator in policy direction, the professor needs to turn from elaboration of his theoretical generalizations under controlled conditions to the use of his ideas in making critical distinctions among conditions in the concrete situation. This shift in the purpose of a course for students of business administration was emphasized for the Program by Harvard Business School's Associate Dean Lombard. He helped to clarify this shift by asserting, "The proper role of theory is to direct inquiry," a significant contrast to the idea of theory as general propositions from which to deduce by logic, action in a particular situation. In a little known book by a professional administrator, a compatible point of view was put in operational steps. Mr. Williamson[2] found, in his systematic collection of questions which executives asked, that getting things done begins with "exposing conditions." To convert statements about things in general to a pattern of inquiry about conditions in a specific enterprise takes not only a different point of view but also a different kind of skill.

Skill: Situational Thinking and Logical Inference

Situational thinking is different enough from the typical scholarly isolation and manipulation of ideas to require much more elaboration than is possible in this book. Much of the case method program in fact consisted of exposuring participants to what was variously called case analysis or situational thinking. This kind of thinking does not get much attention in other fields. A scholar from a traditional academic discipline considered that participants in Harvard's Advanced Management Program for senior executives were "slow to be excited by ideas." He acknowledged that these were intelligent men, able to get a great deal out of case situations, but that the experienced executive, unlike the academician, does not have a bibliography of active analytical ideas in his mind. Professor Paul Lazarsfeld, for instance,

[2] *An Executive Operations Technique*, by C. Don Williamson (Englewood Cliffs, New Jersey: Prentice-Hall, Inc., 1963).

who met with the Visiting Professors in 1960, had an amazing capacity to recall "the analytical concepts" in a seemingly inexhaustible list of books, with their dates and publishers as well as the qualifications of the authors. This total recall could open up a wide horizon of stimulating implications for social structure — if an administrator just knew what he wanted to do, or where he wanted the organization to go. This abstract generalization is not only a different type of generalization from knowing the policy direction of a firm, but it is the consequence of a different skill. Clearly, in today's complex world, the business administrator needs to consider how men with such skill and knowledge can be utilized. Similarly, scholars with such knowledge and ability who have concern for action on their ideas in concrete situations need to appreciate the skill of situational thinking required to create and maintain a harmonious pattern of policies for the direction of a specific enterprise. In the concrete situation, the administrator has responsibility for the totality of the living enterprise in its environment. He is concerned with all of the activities and interactions in the enterprise, with the way people think and feel about them, and the norms and standards involved. He is much concerned with routines and how to maintain access to the exceptions which might indicate the need for deliberate changes in direction because of new opportunities in the environment or needs of a maturing organization.

Maturity: Growth in Balance of Action and Reflection

A fourth objective, growth toward maturity or personal identity, obviously is of concern in all education. Emotional maturity is especially important for those who must take leadership responsibility. As development toward this objective is so central to classroom discussion of cases, it will be described more fully in Chapter V. The differential between the rate of intellectual comprehension and the rate of emotional assimilation and maturity is all too often overlooked in designing a course. The desire to maximize the contribution of each course to all these objectives in the limited time available motivates case-course

EXHIBIT IV-3

THE CASE METHOD OF STUDYING ADMINISTRATION

Bringing Together Multiple Objectives in Case-Course Development

For a Unique Enterprise in its Environment		Conjugate Objectives Both —— And ——		For an Abstract System of Knowledge under Controlled Conditions
Policy Direction	⇄	Generalization	⇄	Pattern of Inquiry
		⇅		
Situational Thinking	⇄	Skill	⇄	Logical Inference
		⇅		
"Can Do" Acceptance — Belonging in the Enterprise	⇄	Professional Orientation	⇄	"Can Learn" Acceptance — Belonging in the Field of Knowledge
		⇅		
		Growth in Personal Identity		
(Impulsive Reaction)	⇄	(Pathology)	⇄	(Obsessive Inaction)
Responsible Participation in Community of Enterprise	⇄	Maturity	⇄	Responsible Participation in Community of Scholars

development. Clearly in a school of business administration all four of these objectives (professional orientation, generalization, skill, and maturity) have a place.

It might be useful to visualize the task of case-course development as one of bringing together all these multiple objectives (see Exhibit IV-3). It would be easier, but unrealistic, to leave on one side the administrator and his specific enterprises and to isolate on the other side the specialist and his subject. However, modern life requires that the student who wants to become an effective administrator and the professor who wants to improve administration must find some way of bringing together increasingly elaborate areas of specialization, because of the emerging needs of unique, specific enterprises, and the growth of the individual.

Sharing objectives, at least temporarily, made participants willing to experiment in case-course development projects, but did not remove that formidable barrier of mastering the number of cases an instructor needed to cover all the courses he must offer. There was an important distinction between the obvious and the operative differences of a typical participant versus a veteran in case-course development.

The obvious difference between the visiting professor considering the development of his course by cases and the case-course veteran was the repertory or ready reserve of cases which the veteran had analyzed and could call upon for use anywhere he thought these cases would be useful for his students.

The operative difference was the willingness of the veteran to make his course an instrument of learning for himself as well as his students — to think of his course as a stage in the continuous process of approximation between an abstract system of concepts and the evolution of an enterprise in its environment by decisions made in concrete situations of things, people, and events.

To be clear about what it means to develop his own course, not even the most prestigious professor in the most prestigious university completely originates his own course. In the words of Erich Fromm: "This is what it means to be original. Not primarily to discover something new, but to experience in such a way that the experience originates in me."[3] Whether the professor is using a textbook, or paraphrasing many textbooks into a lecture series, or using cases, the important ingredient he is trying to recreate with his students is an appreciation that a concept is an invention and that they too can have the experience of making this idea, this body of data, their own. Obviously there is not time to discover the wheel again and all the other concepts which make for modern living. Nevertheless, it is imperative that the professor and his students experience these concepts as their own, as tools to fashion and use for creative action in concrete situations.

Although professors frequently talked about teaching their own courses, they were surprised at the effort required when the Program asked them to reorganize a course with a carefully sequenced series of *decision points,* cases in which their subject was pertinent. This frequently meant that the professors had to revise their concepts to meet the realities in such decision-making situations.

The question of where to start then had to be answered in terms of a professor's making a beginning from where he was rather than to wait until he had achieved mastery. The men who made most progress in case-course development were those who were able to tolerate gaps between the existing structure of a course and the reactions of students to a case introduced in it. But they were not willing to tolerate the gap indefinitely, and sought to close it by getting additional cases and, if need be, by making changes in their conceptual structure. An illustration was given by Professor Virgil Boyd who had been a participant in the 1957 Program and who came back to visit the Program in 1965. Professor Virgil Boyd, then Associate Dean at Northwestern University School of Business Administration, asked the visiting professors in 1965 to discuss a case which he wished them to place in a course in controllership. He pointed out that this matter of sequence might involve change in the conceptual structure of a man's own course.

[3] Erich Fromm in "The Creative Attitude," in *Creativity and Its Cultivation,* Harold H. Anderson, Editor (New York: Harper & Brothers, 1959), p. 50).

Professor Boyd endeavored to help the participants see the issue in the case and what considerations were strategic in developing a plan of action. Then he concentrated on getting participants to think over the accounting courses, which practically all of them had had, to find a place in which the course concepts and the case considerations would be relevant. He explained that he had come to the Program in 1957 as an "old line accountant." Upon his return to Northwestern University, he was asked by the industrialist in this case for advice on how much a particular part cost and at what price it should be sold. As he worked on this consulting assignment, he began to realize that this question did not fit into his accounting course at any particular point as it was then organized. "Here is a case clearly in my field, where does my course deal with it?" This bothered him enough so that he and his associates began to reorganize their accounting courses around such action points. As their students began to think of accounting systems through the study of situations where decisions would make a difference, they not only developed better judgment but assimilated the background information needed for CPA exams so well that they were in great demand by public accounting firms.

Closing the gap between the cases as they are introduced into a course and the existing conceptual structure surely is helped to a large extent by the willingness and ability of a professor to go out and get primary data from actual administrative situations. By such discovery he can see for himself the conditions under which professional orientation as an administrator and as a specialist interact. He can appreciate the connection between the generalization of a policy direction for the enterprise and the clarification of a particular subject. Identifying these interactions also involves increased skill in situational thinking and maturity. Hyphenating *case-course* development is one means to this end, but it is more easily said than done.

Many different types of courses properly belong in a curriculum designed for different groups of students, but objectives must be carefully considered. Surely, if the *purpose is only to communicate a body of information,* a well-organized text or program of learning may well be the most efficient method. If the purpose, however, is to develop *capacity to use information at the point of decision,* the sequential acquisition of knowledge must be converted to make it available for random access at the moment of need. Students who have had to reach out for information in arriving at decision after decision in a series of cases have been able to recall and utilize the information effectively when applicable in new situations or to seek new information when it became relevant.

As information becomes obsolete more and more quickly, the limitation of sequential coverage as an objective becomes more serious. Many of the visiting professors had been searching for an alternative and welcomed the experiments with case-course development as a way of giving students the experience of discovery and random access to analytical concepts. They found that they did not need to abandon the idea of "making a course one's own instrument of learning" even if they were responsible for many courses. Rather, such multiple responsibilities made it all the more important to understand the process explicitly in order better to select and assimilate cases and concepts from others.

The use of cases as a means of helping professors and students achieve such objectives also makes possible this shift in *sequence.* Preoccupation with a logical beginning, progression, and end to a course can be relaxed in order to make the course part of the continuing life of the student, to emphasize the continuing stress between the concrete and the abstract, and to encourage the view that education is a succession of approximations.

The professor may start this process of successive approximations by converting his topic outline into matrixes of concept and context in which his growing reserve of cases could be set naturally. Or the addition of cases might modify the matrix. If the professor is trying to get his students to think of his course as such a continuing process of interaction between concept and cases, he can afford to let his students share with him the struggle to rearrange topics or cases as research and class discussion create new meanings or questions. It thus is not necessary to become the infallible master of all the cases in an ultimate statement of course concepts in order to get started. If such an expectation of shared discovery in the search for common objectives could be established, a description of some experiences in the Program's case-course development might provide stimulus. The faculty who were experienced in case-course development did vary widely in their practice, as shown year after year by visitors to the Program.

Some of the most common operational ideas for clarifying the interaction between cases and concepts are summarized in the next few paragraphs. Each professor would describe his approach in somewhat different terms and conversely his views would be manifest by somewhat different behavior. In Section Two of this chapter, samples from the work on case-course development in the Program are given in enough detail to suggest the interrelationships among these practices in relating concepts and cases:

(1) Building a ready reserve or repertory of cases
(2) Converting a topic outline to a matrix for case analysis
(3) The case-concept-case sandwich
(4) Currently useful distinctions
(5) Course concepts as the magnetic field for case discussion
(6) Formulating concepts from case discussion

Building the Ready Reserve or Repertory of Cases

No amount of general knowledge about business administration or of a specialized subject will automatically provide mastery of the facts in a particular case. Moreover, on the first day of each Program many participating professors were not acquainted with the facts in any case. The idea of building a course out of cases therefore was hard to visualize. But by the end of the Program participants had notes and detailed acquaintance with 40 or more cases. This ready reserve or repertory permitted them to project different sequences.

As each summer went on, informal discussion with experienced case-course designers elicited suggestions for keeping folders of notes about individual cases, their analysis, ideas for use of blackboard or other visual aids, and questions. (Popular casebooks provide instructors with a "manual" of such guides to individual case use.) From year to year a man's folders enlarge with ideas from reactions of colleagues also using cases and from successive groups of students. Fortunately the Intercollegiate Case Clearing House (see Chapter VII) now provides a listing of abstracts of new cases and sample copies for inspection. There are many aids to building a repertory of cases; probably none more powerful than the need for a case for a particular spot in a course matrix or outline.

Converting a Topic Outline to a Matrix for Case Analysis

In discussing field work, the Program suggested that the participant carefully record what he saw and heard, then develop work sheets to help him discover the meaning of what people did, with whom they did it, and the value to them of this experience. Similarly in case writing, the Program suggested that the participant carefully present what he saw and heard — without comment. The Program also suggested, however, that comments be separately recorded in an analysis of the case and as a guide to teaching the case.

Course development involves a somewhat similar dual process: (1) the building of a ready reserve of cases from a person's own field work or from other people's, with careful analyses and notes for teaching, and (2) at the same time, the construction of tentative cross tabulations or matrixes of concepts and analyses used in different contexts in the course.

The idea of such a matrix helped to demonstrate to many professors that their responsibility for coverage of a subject could be met by using cases. It allowed a professor to suspend his concern about sequence until his concern about coverage had been met. A veteran case-course professor in marketing, for instance, would explain that during the term he would select cases that involved students in product selection, segmentation of the market, channels of distribution, etc., and that these cases would also alert students to the impact of consumer goods as opposed to basic raw materials, and to the impact of distribution logistics. Or in labor relations, a professor might have across the top of his tabulation such topics as organization drives, contract negotiation, grievance procedure, etc., while vertically he would be concerned with skilled crafts, industrial unions, government workers, etc. The professor is constantly tightening his matrix and seeking cases which cover more cells of the matrix so that his course will make the greatest possible impact in the few class periods available.

Unquestionably the conversion of an existing course to the use of cases would require a considerable repertory of cases. It is unlikely that anyone would make a complete switch at one moment, although logically this might be easier in the sense that the professor would see how the topics of his course would eventually be covered in the total sequence of cases. Introduction of one or a few cases involves something of a paradox. Consideration of each case draws potentially upon the entire subject. Or, in current terms the case requires a "random access program" to the concepts of the course. At the same time, the careful analysis of a case is time-consuming, and an instructor new to cases may feel that he will sacrifice the coverage of some points in his course until he sees how cases overlap to fill the cells in his cross tabulation of concepts and contexts.

Chapter V dwells more upon the classroom discussion of cases and the possible advantages to student learning if the professor can relax concern for logical topic sequence and involve his students with him in looking at cases as an opportunity for mutual discovery. Thus, neither the professor nor the students need to wait until the end of the course to begin to use the conceptual system of the course from case to case.

The experiments in the Program revealed some of the many ways in which different faculty members dealt with this persistent stress between coverage of the subject and progress toward the multiple objectives of education for administration by sequencing a series of cases and concepts.

The Case-Concept-Case Sandwich

One of the most frequently used devices is the "sandwiching" of cases and concepts. Cases are selected by the professor and interspersed with technical notes or research findings which put concepts of the course before the students. Sometimes professors will devote a classroom period to discussion of such research findings or concepts. Frequently, to save classroom time, readings are sandwiched into the assignments given to students. How they are then utilized in case discussion is a subtle matter of moving easily back and forth between the world of the business administrator and the world of the academic scholar. Illustrations are given, in Section Two of this chapter, of comments by participants in the 1959 Program, namely, "We were introduced to an increasingly important field which derives much of its theoretical content from the behavioral sciences." Whatever this process of derivation might have been, all too little of it is being done. Perhaps it may be exponentially more difficult than generalization about the determinants in a single value system. In any event, this process is critically important in the development of case-courses in business administration, but it certainly is not accomplished without effort.

Currently Useful Distinctions

In interspersing research findings and conceptual statements in the series of cases assigned, it is difficult not to divert attention away from cases toward knowledge in general instead of getting students to see the interrelationship. The skilled case discussion leader uses knowledge, uniformities in relationships which researchers have found as a means to help students see more precisely the situation confronted in a particular case. He introduces readings into the case discussion, in terms of the case itself, by using the distinctions which the researchers implied. Note, for instance, in Section Two of this chapter, the way the instructor encouraged participants to make the distinction between job satisfaction and productivity under different conditions of social organization.

The Course Concepts as the Magnetic Field for Case Discussion

In the sessions led by Professor Lawrence, reported later in this chapter, it will be evident that much of the discussion came out in whatever way participants found it comfortable and convenient to voice their own ideas. These comments were haphazard and random in terms of Professor Lawrence's course concepts, but by utilizing the blackboard he was often able to arrange these random contributions in order to reinforce the distinctions he felt the course concepts made in analysis of a concrete situation. His behavior in the classroom sessions further reinforced the insights students obtained from assigned readings.

This conversion of research uniformities into penetrating tools for analysis of a concrete situation is at the crux of case-course development. Having made this conversion, first as part of his field case research, the case-course instructor then consistently follows through in class to help the participants use these distinctions in discovering a new meaning in the specific situation. This is the skill of situational thinking. By his questioning, the instructor further orients participants to the administrative role and the use of enlarged insight for discovering alternative courses of action. He is even able to get participants to take a position, to become aware of their own feelings and beliefs against reality requiring action.

Alumni of case-course programs have recalled vividly how different faculty members with different personal styles have maintained a pattern of discussion. They remembered with some pain the analogy of the old physics experiment: When one holds a piece of iron in a magnetic field and hammers on it, the molecules will align themselves in the field until the bar has itself become magnetized. Many coaches expressed regret over the years that the Program felt constrained to operate in such a way that participants could not be held more firmly under such consistent hammering. The relationships with visiting peers was such that the strain of this discipline was more than a group with heterogeneous motivation found comfortable for more than a demonstration. It takes getting used to. The idea that knowledge is something to be used or the possibility that knowledge might actually be different for different purposes appeared to be unsettling. The Harvard coaches had not realized the unusual extent to which, in the Harvard Business School programs, 6 or 7 sections in a class of 650 men encouraged instructors to discuss individual cases and course objectives, and to argue over concepts, research findings, information, or anything else which would assure penetrating analysis and appropriate action in a case.

Formulating Concepts from Case Discussion

Another practice as illustrated in Section Two of this chapter is to encourage free discussion of two or three cases and *then* to formulate with the students the relationships which had come out as important considerations during the discussions. It is one thing to describe this process and quite another thing to experience it. Professors were prone to rush conceptualization. The Program provided evidence that students, even professors as students, were not ready to see meaningful conceptual relationships without a backlog of shared case discussion. In 1964, for instance, the group of participants with Professor Walter Salmon as coach led discussions of about a half-dozen cases involving the creation of market demand. At the conclusion of these discussions, Professor Salmon's success in coalescing the group's collective exposure to cases involving emotionally significant purchases and visibility of product quality was extremely valuable for all participants in the 1964 Program. In another effort to accelerate the process of case-course development, Professor Salmon agreed to start the 1965 Program with this same sort of sequence, to show how formulating general relationships in marketing was related to reaching a policy decision in a particular enterprise. In 1964, however, before Professor Salmon's brief demonstration, the visiting professors had already become a working group with a backlog of at least two dozen discussions of varied cases and four weeks of field research and case writing. In 1965 Professor Salmon was asked to try to demonstrate this balance of policy action and subject generalization the day the visiting professors started. Although he was even more brilliant than in his impromptu synthesis given the previous year, it was too early. The participants had not yet done enough work together; they were not ready for such generalization.

SECTION TWO

A Sampling of Case-Course Development Efforts

The first section of this chapter closed with brief mention of some pragmatic approaches to the task of relating cases and concepts in a course. The following sampling of experiments from the Program will show the mixture of efforts used and the way in which responses of participants in turn were affected by the way they saw these efforts in terms of their own purposes. The samples were chosen partly because these were the records that were available. This was not a research project but an operating, cooperative evolution. Also, the development of a case-course cannot be considered apart from how the cases might be used in the classroom. Hence, in the final episode reported, considerable attention has been given to detailed classroom reactions to cases and concepts.

Concentration on the Policy Direction of an Enterprise, 1957

The first two years of the Visiting Professors Case Method Program, summers of 1955 and 1956, relied heavily upon selective observation of the Middle Management courses by small groups of visiting professors. The Harvard Faculty members teaching different courses met with the visitors and explained course objectives and plans for each day's class. Visiting professors would observe and then have another opportunity to discuss with the instructor discrepancies between objectives and what in fact took place. Valuable as these visits were, it was evident to all concerned that they did not take the place of actually experiencing a course developed through the use of cases.

In 1957, Professor Franklin Folts and his colleagues who were teaching in the Middle Management Program at the Harvard Business School felt that, judging from the previous summers' experience, the participants in this Program would not be disturbed if there were more active participation by the visiting professors in at least the Business Policy course. Professor Robert

Merry, who was to teach the course, had long been interested in the whole teaching-learning process and had worked with the predecessor program for Harvard's own new Faculty members. By happy coincidence, in 1957 the Business Policy course was to begin in the Middle Management Program soon after the Visiting Professors arrived. The Middle Management Program at that time extended for eight months and had begun in January. The students and Faculty, therefore, had developed considerable rapport before entering the final stage of the session which overlapped the schedule of the Case Method Program.

On the opening night of the Case Method Program, Professor Merry outlined his objectives for the course and the way in which he expected the visiting professors to participate fully in this one course within the Middle Management Program. Inclusion of the visiting professors was not the only innovation, however; Professor Merry that year was conducting all three of the weekly class periods devoted to the Business Policy course on a single day, one period right after the other each Wednesday. By this course concentration, he hoped to create a more realistic experience showing how one decision in a business situation moved into the need for another decision and so on. Professor Merry announced that in the Middle Management classroom he would expect to treat the visiting professors just as he did the Middle Management men, although he did not intend to call upon them unless they wanted to volunteer. Professor Merry went on to explain his willingness and expectation to meet separately with the visiting professors from time to time to discuss what was going on in the course. He would then listen to questions and appreciate any reactions concerning his conduct of the class.

In one of these early pedagogical review sessions, Professor Merry illustrated his objectives for the course by referring to the discussion of the Castengo Steel Company case. He said that he wanted the Middle Management participants to see that it was not necessary for them to accept a business as it was. He felt that Middle Management men tended to assume too often that what a company did was good. He felt the course should give men more vision, should make them see a wider range of possibilities in a situation, and should help them increase their willingness to change the business. One of the visiting professors felt that, for the participants to do what Professor Merry wanted them to do, they would need to know much more about the industry. Others were less concerned about background industry information but more concerned about the questions the professor asked, both in preparation before class and in participation during class. Professor Merry acknowledged that if an instructor wanted to be popular in the short run, he might let students talk about what interested them and even lecture about the questions he had asked before class, but that to be effective, the case method instructor must drive student thinking further than the students had gone by themselves.

Professor Merry noted that more structure was required when the week's three class discussions were held in sequence on the same day. Professor Merry felt it was important to make participants present their programs of action in sufficient detail so that they could be considered in an orderly way. Too often students would merely go down the notes they had made rather than respond to the specific question he was asking in order to get analysis on a systematic basis. It was apparent that, although the visiting professors had been conscientiously preparing the cases for discussion in this Policy course, they felt restive in the classroom because Professor Merry was taking up the discussion of the case in a way which they had not anticipated. They did see that Professor Merry was seeking to get orderly subdivisions of a complex situation in which the executive could exercise judgment in relating and combining the facts. He pointed out that he felt it was especially important to do this with the Middle Management men because they were prone to jump to conclusions for action based on hunches about the total situation, without genuinely considering components of the situation. He acknowledged that the results were not so much a matter of how the questions were formulated as the kind of relationship which the instructor could develop and maintain with the class.

The original intent had been to give the visiting professors experience in working through a live course as it was being developed through cases, but the plan had to be modified for a variety of reasons. An important reason, and perhaps the most acceptable one, was the need for participants who were writing cases to visit the companies when the companies wanted them to come. This priority made for some irregularity in attendance at the Business Policy classes. Also, the visiting professors hesitated to participate in the class discussion and, as observers, were not getting the full emotional impact of the course. As hosts to the visiting firemen, the Business School could not overly insist upon attendance in the Business Policy course, although increasing absenteeism provided opportunity for the group to discuss how much of the value of an educational program comes from the pressures which provide a person, even a professor, with the impetus to do what he does not want to do.

In the discussions with Professor Merry and the visiting professors about what was taking place in the Business Policy course, there were unusual opportunities to see how a number of elements became involved in the development of a course. The tandem sessions, at least for the visiting professors, seemed to lose interest rather than to exploit the possibilities in each case. The cases had not been written, nor the field situations studied, with this classroom sequence in mind. In Chapter III, Professor McNair discussed the time structure around which a case is written and the troublesome questions of moving from the duration of months or years in the field situation to a single classroom period. Professor Merry was experimenting with restoring more duration to the discussion. The experiment demonstrated that intensity of concentration does not fully take the place of reflection and subconscious association over time. During the summer program, everyone noted that having an adequate interval in a course to sleep over a case discussion before going into the next one did have an important place in the learning process. Time for mulling over a person's own participation and that of others, as well as the instructor's questions, is an ingredient in the impact of a course which is accentuated by the *duration* as well as the *intensity* of the experience. With such tandem sessions in mind, subsequently members of the Faculty have done a great deal more in the development of series cases. In these, a sequence of experiences in time, or a variety of experiences in different phases of the enterprise, could be dealt with independently but still as a whole. (See for example Professor John Glover's presentation in Appendix 4.)

The visiting professors did see that embarking on the development of a course through the use of cases brought to the surface complexities which, although they might exist in a series of lectures or chapters of text, did not become quite so obvious and importunate. The effect of student preparation clearly limited what could be done in carrying out a professor's design for a sequence of classes. It became clear also that the effectiveness of preparation was not entirely a matter of effort. It was also the product of skill in analysis. Small group meetings before class brought out a significant contrast between the professional orientation of the professors and that of the management participants. The professors typically resisted commitment of so much time to the problems of an enterprise in which they felt no continuing interest. The Middle Management men, however, found realism in the small group meetings and an increasing awareness of Professor Merry's purpose — getting them to think of a company as something which the administrator could run rather than be run by.

The experience with Professor Merry in the Business Policy course gave participants intensive immersion in the facts of a few enterprises and the interrelationship of these facts. Professor Merry was concerned with the impulsive willingness of the Middle Management participants to act without thorough analysis and their passive acceptance of company policy. He, therefore, had designed the Policy course to meet these conditions. The visiting professors, although they could appreciate these objectives, did not have sufficient commitment to understand the interrelationship of facts in one case situation in order to enjoy the full day of discussing a single case.

Furthermore, the visiting professors generally did not see the relevance between their course responsibilities and the policy decisions in which Professor Merry was involved. Even those professors who were to teach Business Policy saw an impossible gap between their own student situation and the intensive program designed by Professor Merry for the Middle Management

EXHIBIT IV-4

Major Considerations Interacting in Course Developments[a]

PROPOSED GOALS (desired changes in students)
- *Knowledge*, to be learned
- *Skills*, to be acquired
- *Attitudes*, to be developed

CONSTRAINTS
- STUDENT MIX
  - *Education*
  - *Experience*
  - *Ability*
- TIME AVAILABLE
  - *Number of sessions*
  - *Length of class period*
  - *Preparation time* from students
- BUDGET
  - For the *course*
  - For *facilities*
  - For *other purposes*
- MATERIALS LIMITATIONS
  - By *administration*
  - Of *facilities*
  - On *total price*
- ENVIRONMENT (attitude)
  - Of *chairmen*
  - Of *faculty*
  - Of *administration*

SELECTION OF COMPONENTS (priorities, trade-offs)
- *Goals*
- *Subject matter* units
- *Achievement*

SEQUENCING OF COMPONENTS
- *Subject matter* logic
- *Difficulty of learning task*
- *Prerequisite understandings*
  - *Skills*
  - *Influence of preceding class*

REVISED PURPOSES: GOALS
- *Knowledge*
- *Skills*
- *Attitudes*

TEACHING THE COURSE
- Effectiveness of the *instructor*
- *Informal organization of the class*
- *External influences*, as draft

EVALUATION OF THE COURSE
- *Students* reactions
- *Instructor* satisfaction
- *Evaluation methods*

COMPONENTS AVAILABLE

- *Topics*
- *Goals*
- *Method and Materials*
- *Achievement desired*
  - Awareness — Familiarity
  - Understanding — Proficiency
- *Acceptable dispersion* of class
- *Learning time*

THE PROCESS — Course Evolution

- Continuous flow and much back-and-forth
- *Goals,* reconsidered
- *Constraints,* changed
- *Components,* newly available, traded or resequenced
- *Prerequisites,* revised
- *Pedagogical* experience
- *Goals,* re-revised
- And so on

[a] Outlined by Robert W. Merry, Visiting Professors Case Method Program, Harvard, 1965.

Program participants. Frequently the Business Policy course in other schools might be the only one in which students were exposed to case discussion. Therefore, under these circumstances, participants began to question the appropriateness of cases designed for use with experienced executives or students who had already studied 500 cases.

In retrospect, the visiting professors were being asked to go along in an intensive program designed to develop responsible active administrators. The assumption that this would thereby make the professors responsible, active, case-course designers had some significant limitations. As mentioned earlier, Professor Walker's advice also had merit for the design of the Program itself: "Keep the beacon light of purpose before you at all times."

Professor Merry was one of the coaches for the 1965 Program, and prepared a comprehensive outline of the whole process of course development.[4] Discussion of this outline was one of the highlights of the Program.

> The central issue in course development is "how does an instructor wish his students to be different because of having taken his course?" Necessarily, therefore, the beginning of all course development should be the establishment of goals in terms of the student, not in terms of the teacher, nor of the subject matter. The practice of treating what is taught or what is assigned as equivalent with student learning is extraordinarily prevalent in higher education despite its widely known fallaciousness. Effective course development avoids this pitfall.
>
> Effective course development involves selection of methods for particular purposes that within the resources and capacities of the institution and the instructor will most facilitate the learning experience on the part of the student. Hence, any attempt to determine in advance that a course will be a "lecture course," a "discussion course," or a "case course" is ordinarily undesirable.

Professor Merry's outline also helped to put into place other experiments to involve visiting professors in case-course develop-

[4] "Course Development," by Robert W. Merry. The substance of Professor Merry's talk developing this outline appears as a chapter in a book by Professor Roy Penchansky entitled *Health Services Administration: Policy, Cases and the Case Method* (Cambridge, Mass.: Harvard University Press, 1968).

ment. Since it always was difficult for visitors to experience that accumulation of understanding which comes through mastery of one case after another in a course sequence, and since the task depends a great deal upon the handling of facts and reactions to them in class, another experiment in the Program will be explained in detail later on in this chapter.

Visiting Professors Give Variety of Short Courses

The principal sample of the case-course development activity during the first three sessions of the Case Method Program, as has been noted, had been the observation, participation, and discussion of courses within the Middle Management Program. Thus, termination of that program caused much concern. Foreseeing the end of the Middle Management Program, participants in the summer of 1958 discussed the possible effect on the case method program. Alumni of earlier Case Method Programs also were asked for their reactions and suggestions for alternative activities. Their primary concern was that an alternative way be found to give participants a sampling of how a course was developed over a period of time through the use of cases.

Many helpful suggestions were received, including the much-appreciated opportunity which a former visiting professor, Professor David Carson (and his Dean, Philip Ragan of Boston University), provided by allowing participants to observe undergraduate classes in business administration in which cases were being discussed at Boston University Summer School. There was much favorable response to the idea of getting participants to prepare a course outline and select cases for it. Some added that it could "by no means be considered a substitute for observation of the Middle Management course." One of these participants added significantly that it was only "*later* (after much case discussion and observation) that I became attracted to the problem of how the courses had been set up." This comment helped in seeing the importance of building a repertory as a first step in course design.

Many alumni also insisted that participants should be given the opportunity to observe a professor who was experienced in the development of a course and who worked from case to case with a group. In the absence of any other students, alumni of the Visiting Professors Case Method Programs urged a revision of the role of host and suggested adding a course in which a Harvard professor would teach a series of cases to the participants. This was finally tried in three different programs. Before analyzing one of these series, a selection of the short case-courses which participants arranged for themselves will be reviewed.

The professor who develops his own course has a major advantage over the visiting professor who was participating as one of a group of five or so others collaborating in the conduct of these short courses. Without criticising the participants, it might be noted that it is well nigh impossible for five men who have been strangers to each other, and who might share little professional interest in the subject, even to approximate holding the discussion of a subject to a common objective over five class periods spread out over two weeks. It took special effort, therefore, to reveal the contribution of these sequences to case-course development, although they clearly did give participants an opportunity to try out various styles of case discussion leadership.

On "Inventory Management"

In 1960, one of these groups of participants, of which Professor Robert McGarrah was coach, developed a short course around the subject of inventory management. Professor McGarrah had selected in advance a panel of cases relevant to this subject. Upon arrival at Harvard, the visiting professors looked over the panel and, among themselves, picked five cases which they felt could be taught as a sequence to give the rest of the participants a better understanding of the administrative problems in inventory control. In this sequence one of the cases rather clearly brought into the discussion the problem of getting an estimate of sales, the cost of setting up the equipment to run the product, the carrying charges of inventory, and the unit cost of the operation. Another case in the series added to these considerations the organizational relationship of the staff production control men and the line foremen. Another case introduced the problem of the company's financial condition. Still another case in the series involved a long-term effort to stabilize employment.

During this sequence of case discussions, one of the visiting professors, James Logan of Columbia University who had led one of the case discussions, explained how at about the mid-point in the series he would have helped his students recognize that they had been struggling to formulate the relationships among variables which recurred repeatedly in these problems of determining economic lot size. Then, in no more than five minutes, he showed how economic lot size is a function of the total volume to be produced, the cost of setting up the operation, and an inverse function of the costs of carrying inventory. With the equation thus stated, he went through the differential calculus analysis to get a volume which would minimize operating costs and carrying charges of the finished inventory for different set-up costs to arrive at the classic formula. Participants, having been alerted to the uncertainties in these various estimates and the intangible implications of the decision from previous case discussions, were in a position to use the formulation as shorthand for getting quickly to a number which might then be qualified by the other considerations in the specific case. All were impressed with the way in which Professor Logan had fixed and made more usable the relationships in this category of case situations by identifying key relationships with a mathematical model. Note also that most of these professors had "known" the economic lot size formula from their earlier academic work but had not assimilated it as a conceptual tool until this five-minute lecture amid five case discussions.

On "Creating Demand"

In 1964 one of the groups, of which Professor Walter Salmon was coach, was developing a short course using cases involving advertising policy. The cases covered a wide range of products from petroleum, Volkswagen sales and repair service, to cosmetics and patent medicine. When some of the participants were protesting after several cases that there was nothing to marketing and advertising but "brass," Professor Salmon did a ten-minute run-down of the case series. Suddenly there emerged from the blurred recollections of this case series a pattern of relationships which put together the company's need to advertise as varying directly with the emotional significance of the product to the consumer and inversely with the visibility of the product quality.

On "Different Costs for Different Purposes"

The significance of the way in which one's thinking is programmed appeared even more clearly in the discussions of short case-courses dealing with accounting problems. From year to year the sequence of cases might be called something like *Different Costs for Different Purposes* or *Relevant Costs*. This series would usually include one case involving a decision whether to make or buy a part in a manufacturing company, another on whether to cut prices on bidding for an order when there was idle capacity, or an appraisal of departmental operations under different plans of allocating overhead. Again as the case series went on, a few words about the way different costs might be considered for different purposes seemed to make the whole idea of an accounting system come to life as a management tool.

In all these situations there would be a protest, "Why didn't someone just give us a short lecture instead of dragging us through all this discussion?" Then a poll would be taken and, with rare exceptions, everyone in the group would have had at least one collegiate accounting course. Nearly always there were several men in the group who had given lectures on economic lot size or marketing. Knowledge was there. Why was it not available when needed? These experiences illustrated that, as one prepares to face administrative situations, new ways of assimilating knowledge become necessary to make it relevant. The language of the abstract generalization must be translated into concepts which the administrator may use to discriminate among conditions in the concrete system for which he must develop a plan of action.

The experiments of grouping professors for particular topics, and having them collectively select cases for use with these topics, had only limited success. This was in part because the professors did not know one another well, and probably because the difficulty of working as a teaching group was underestimated. Furthermore, the rest of the participants seldom had the disciplined interest which a professor in a specialized course might expect from his students.

Out of such short case-course experiences it became clear that most men simply had not had the opportunity to become acquainted with enough cases. Without a repertoire of cases there was no way of selecting, sequencing, and capitalizing on an emerging awareness of uniformities. As each summer went on and 40 to 75 cases were analyzed, visiting professors could begin to see many more possibilities for using case discussions as building blocks. But by the end of the summer it was too late for the group to start a major project of going back and rearranging cases into a new course framework. Reluctantly, therefore, experiments were tried having a Harvard professor put a course together and leading the visiting professors through it. In 1959 Professor Paul Lawrence, who had been a coach the preceding year, organized a series as part of a major effort to get across the idea of case-course development.

Fortunately, because of concern about the shift in asking the visiting professors to act as students, rather complete notes were kept of this experiment. Some of the detailed class discussions will be reproduced here to show how the analysis of individual cases and the use of research findings and theory may be interwoven. The same notes might also serve to illuminate the conduct of class discussion. The course is developed from case discussion to case discussion. The course objectives likewise determine the direction of class discussion. The context in which the course development project was organized will now be presented.

Although each professor has his own style, among those who use cases to develop a course there are some common processes. The use of time in class when cases are discussed differs enough from a series of lectures that one report in considerable detail has been given. Only by detailed attention to the interaction between instructor and students in relation to data in the case, concepts of the course, and data in additional cases can an instructor begin to see how cases are used to communicate concepts rather than to illustrate them. Professors naturally choose cases and use the considerations which are important in their specialized subjects. The process of interaction between instructor and students around data and concepts is much the same whether the subject be work flow and method analysis in production, distinction of product characteristics and consumer needs in marketing, the flow of funds and time value of money in finance, or the interaction of individual needs and group norms in organizational behavior. The most detailed illustration available from the Program came from a special effort to help participants experience for themselves how one instructor was seeking to relate cases and concepts in work group behavior. Although the materials illustrate the conduct of case discussion, the report also indicates the ratio of cases to readings, even the amount of time taken in class to let participants voice their reactions to the case and when and how within the class period the instructor communicated the conceptual scheme of the course by using it to help organize the response of participants to the case. Again, the detail reproduced here is not given as something to be imitated, but as data to consider in organizing a matrix of concepts and in selection and sequencing of cases in development of a man's own course.

Management in a Social System

When the participants gathered on Monday morning, June 15, 1959, the first order of business after introductions was a discussion of "plans for course development."

Four types of activities were outlined:

(1) There would be observation of regular classes at Boston University and of special management programs on the campus at the Harvard Business School.

(2) Members of the Harvard Faculty would come in from time to time to discuss their individual course development activities. (A sampling of this approach began that morning

when each of the four coaches spoke briefly about his own course interests by way of participation in the general introduction.)

Wilbur England — Procurement
Walter Frese – Financial Accounting
Pearson Hunt — Financial Management
Thomas Kennedy — Industrial Relations

(3) Each participant would develop an outline for a course of his own, drawing upon the freely available cases of the Intercollegiate Case Clearing House. Bibliographies were thereupon distributed and procedures explained for making up lists of samples desired and a target date for the preliminary lists set for ten days later, June 25.

(4) We would all participate with Professor Paul Lawrence for 16 sessions in which he would work with us on his immediate task of merging cases and concepts in one section of his course entitled Administrative Practices. For purposes of the summer program, this activity would be shown on the schedule as "Management in a Social System."

The purpose of this special course to be led by Professor Lawrence was carefully explained so the visitors would take seriously the basic premise that all coaches and participants were sharing together the exploration of the case method. The coaches were not taking the role of instructors and asking participants to be students. The stage was being set for "Paul" Lawrence to be just another professor-participant trying, as each participant was trying, to select and sequence a set of cases to accomplish specific course objectives within an allotted number of class periods.

During the first week there were visits to Boston University classes, discussions of cases led by participants, visits to business firms, analysis of case teaching as it appeared on a filmed discussion, small group discussions of cases, and a large-scale demonstration and analysis of case leadership by Dean Richard B. Donham of Northwestern University.

In the second week an intensive 32-period course began. In alternate periods participants acted as students of administration and in the next period as professors concerned with the relation of cases and concepts being studied. This is the way the schedule appeared:

<u>Monday, June 22</u>

9:00 a.m.	Aldrich 12--"Management in a Social System"--Professor Lawrence #1
10:00 a.m.	Sherman Hall--Coffee--Discussion of "Cases and Concepts"
11:00 a.m.	Aldrich 12--"Management in a Social System"--Professor Lawrence #2
12:30	Kresge Hall--Luncheon and further discussion of "Cases and Concepts"
Afternoon:	Case Research and Selection of Cases for Course Development
Reminder:	Preparation needed for two case discussions on Tuesday

<u>Tuesday, June 23</u> (a repeat of the Monday schedule with one addition):

4:00 p.m.	Aldrich 12--Professor Donald F. Blankertz, Director of the Graduate Division, University of Pennsylvania, Wharton School of Finance and Commerce--"Limitations of the Case Method"

The outline of the cases and other materials Professor Paul Lawrence distributed to the group is shown in Exhibit IV-5. The first four discussions in the schedule will be reviewed in considerable detail and then attention will be given to some of the queries Professor Lawrence and the coaches had about the apparent lack of interest among participants in pursuing the pedagogical questions of merging cases and concepts in course development.

Professor Lawrence began the first class with the visiting professors by giving a 25-minute talk on the background of the course for which he was responsible in the first year of the MBA Program: the Mayo-Roethlisberger research, problems in human relations, management controls, administrative practices. He explained the present focus on organizational behavior. The course objective could now be stated as communication of knowledge about organizational behavior *and* development of "skill in use." The approach was to make students aware of the growing body of concepts, "theories of explanation," the research findings, *and*

EXHIBIT IV-5

HARVARD UNIVERSITY
GRADUATE SCHOOL OF BUSINESS ADMINISTRATION
George F. Baker Foundation

SUMMER CASE SEMINAR PROGRAM

Administrative Practices Sessions

Table of Contents and Course Schedule

Session No. and Date		Cases and Readings
1.	June 22	Dashman Company IM 1356R
2.	June 22	Gordon Foundry Company ICH-2H23(R)
3.	June 23	Martin Company EA-A 11 Selections on "Map-Territory" EA-A 284
4.	June 23	Superior Slate Quarry(A) HP 236A (R) Worker Satisfaction and Development, Zaleznik (Foreword and Introduction)
5.	July 1	Excelsior Bakeries HP 478
6.	July 1	Mallard Airplane Company (A) EA-A 185 "Management and the Assembly Line," Turner Harvard Business Review, Sept.-Oct., 1955
7.	July 2	"Individual Needs and a Need Hierarchy" Barnes EA-A 289
8.	July 2	Merrill Manufacturing Company A, B, C, D, E HP 272 (R)
9.	July 8	Marshall Company M-R3, EA-A 72 M-R3
10.	July 8	Federal Auto Parts Company EA-A 98
11.	July 9	Claremont Instrument Company EA-A 108
12.	July 9	Haig Chemical Company (A) EA-A 34A
13.	July 16	Flint Electric Company A, B EA-A 74R
14.	July 16	Flint Electric Company C EA-A 74R
15.	July 17	"The Foreman, Master and Victim of Double Talk," Roethlisberger, Harvard Business Review, Spring, 1945

cases in which the student might find this knowledge relevant.

While he spoke, Professor Lawrence was sketching on the blackboard the visual relationships as follows:

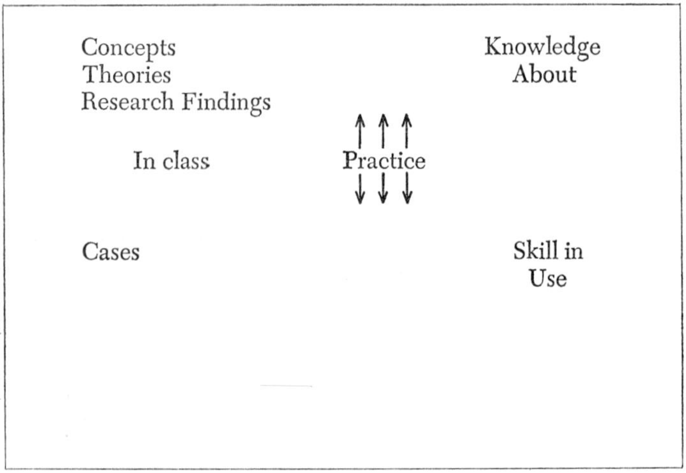

Much of the time in class was to be given to "practice in going back and forth between cases and concepts, between knowledge and skill." This was the plan Professor Lawrence had in mind for the section of the course in which coaches and visitors were going to work together. The *substance* of this section had to do with "WORK GROUP BEHAVIOR IN INDUSTRIAL ORGANIZATIONS." The outline of materials previously distributed included readings which he expected the group (visiting professors and coaches) to study and cases in which all were to use skill in dealing with concrete situations.

After the 25-minute introductory talk, he turned to the Dashman case and asked the group: "What has Post run into?"

> The Dashman Company case is two pages of description about the approach Mr. Post took in late 1940 when he was made vice president in charge of purchasing, a position especially created for him. As a first step he decided that he would require each of the executives who handled purchasing in more than 20 plants in different parts of the country to clear with the head office all purchase contracts which they made in excess of $10,000. During

the next six weeks the head office received no notices from any plant that contracts were being negotiated. Executives in other departments who made frequent trips to the plants reported that the plants were busy, and the usual routines for that time of year were being followed.

A participant volunteered that Post was just a name to the organization. Another participant added an episode from his experience. In the next five minutes several comments were made, which the instructor capsulated without comment and noted on the blackboard.

At the end of five minutes and after seven men had volunteered comments, the board appeared about as follows:

No personal contact	verbal agreement
	passive do nothing
Resistance to change	informal contacts
Threat to Prerogatives	
Loss of power	
threat to job	
App't of Post?	

Without interrupting the flow of comments from the class, the instructor pointed to the second column and said, "Here you are noting *forms of resistance*," and, pointing to the first column, "Here are *the reasons why*."

Other participants presented further reasons: poor timing, fear of the unknown, no reason given for the order, inherent staff-line conflict, relationships of purchasing agents to plant managers. (The instructor added these to the first column.)

After this had gone on another five minutes, the instructor asked: "Did any of you figure out what Post is up to?"

While discussion went on in the next few moments, he went to the board and inserted over the lists which he had previously put on from class comments, the headings:

REASONS, EXPLANATIONS	FORM OF RESISTANCE	
		(This third area of the board had not been filled)

In the next ten minutes there was a very active emphasis and re-emphasis on different explanations and assumptions of the reasons for the resistance, in addition to efforts to fill the void on the board. Soon the group seemed relieved to have a third heading put up:

		PURCHASING SYSTEM

Attention was then extended to speculations about why the $10,000 limit had been set; the extent to which government restrictions were altering procurement patterns; and how allocations of scarce supplies would be made within the company; which discussion, the instructor summarized to the effect, "You see Post as breaking a code in the behavior of staff and line as to giving advice or orders."

"Let's see if we can put up an organization chart."

An hour had gone by at this point and the coffee break was approaching. Men with some experience or recollection of the prewar scarcities were strongly advocating, "pull all the purchasing agents into the home office." They were being drawn out by the instructor in two different directions as he asked, "How would you build a case for this?" or "What is this company's present form of specialization?" This latter line of thought was being sponsored by a participant because he believed each plant had specialized needs.

At about this point Professor Wilbur England, one of the HBS coaches who taught the Procurement course broke his silence and interrupted one of the participants with the exclamation, "I object to men teaching purchasing who do not understand it!" The good-natured comeback from the visiting professor whom England had interrupted was "Since when is the purchasing agent important?"

By this time everyone was talking at once and only partially subsided to let the instructor try to express the companion questions which the group seemed struggling to clarify:

What didn't Post know about organizational behavior which we might have known?
What is a good way to purchase in wartime?

Although the instructor seemed ready to adjourn with this closure, the discussion kept on going as the class broke up and went toward the door. "He should have sold the idea at the top." "He needs etiquette rather than threats." "What do you mean by the weight of the organization?" "He has it already in the Board of Directors.

By design, the participants should now have shifted their attention from the study of organization for wartime procurement to their normal role of professors analyzing another professor's case-course plan and performance.

Professor Lawrence and the Harvard Business School group who planned the summer had anticipated that the visiting professors would find the coffee break pedagogical post-mortem much more interesting than the administrative action case, but the interest of the visiting professors in the job of the executive had been underestimated. The visiting professors had become so much involved with Mr. Post and his boss, President Manson, or his assistant Mr. Larson, or the plant managers or their purchasing agents, that no one was really willing to sit back and consider what the instructor had been doing during the preceding hour. Although the group, coffee cups in hand, finally assembled around a big table and tried to give Professor Lawrence a chance to say what he had been trying to do, the positions taken in the class discussion were still actively at work and brought the discussion back to Mr. Post.

Before that morning, concern had been how to get professors into a case discussion, especially those professors who had been asking how knowledge and concepts could ever be communicated by use of cases. Now the problem was to pry them from the case long enough to listen to the instructor explain what knowledge and concepts he was utilizing in his course.

The first case at least demonstrated that even professors as participants saw quite different things in the same two-page case — a surprising discovery for many who after two hours still could not believe that their colleagues really read such different meanings into such an "obviously" clear and simple statement.

Maybe there was an emerging awareness that those two pages of data selected from the field situation by the case writer constituted a different kind of material from two pages of a textbook. There could have been a sense that, in assigning those two pages for study, the instructor had something different in mind from the sequence of ideas from chapter to chapter in an exposition of a subject.

After the coffee break and the vain attempt to shift to the pedagogical issues, Professor Lawrence did spend a few minutes on a brief preview of the next case and what he was trying to do in discussing it. There had been considerable discussion around the coffee table about the extent to which participants felt the instructor had "controlled" the discussion. All had felt free to bring out what they wanted to and to voice their beliefs about the meaning of the situation, but some noted that the way the instructor had used the blackboard organized if not controlled what was being volunteered and the direction of the discussion. The relationship between his organization of comments and the concepts of the course had not yet been considered.

The second case was a one-page case from the University of Alberta, GORDON FOUNDRY CO. In it a graduate from the Provincial Technical Institute tells of an episode that occurred about a month after he went to work as the assistant to the president.

Professor Lawrence opened the period by asking, "What is going on here? What is he running into?"

> For 15 minutes comments streamed out, while a word or so was jotted on the board which seemed to catch the thrust of each man's thought. At that point a participant came up with a statement to the effect, "He should sell himself by seeking the help of the older workers." To which the instructor replied, "You have made an action proposal from a summary that it is a personality problem." The board at that point looked about as follows:

John	Group	Motives	Action Proposals
engineer	older	job security	sell himself
young	long service	solve difficult tech problems	seek others' help
technically trained	close knit		
	know-how		
	pride in workmanship		

For the next one-half hour the discussion flowed on around questions which the instructor phrased in terms which had been volunteered from the class. "What the eager beaver was trying to prove," "What kind of relationship to the group he needed to develop." The instructor had interposed at one point to ask with respect to one ingenious explanation, "When was the technical assistant hit by the hose?" "Was it while he was on the job?" "How was he interpreting cat calls?" "What is he spending his time thinking about?"

By this time 45 minutes of the period had passed and all but two of the participants had become vocally involved. The instructor then asked, "Where are we?"

While the group was taking stock, he circled the comments on the board about the technical assistant and made another circle around comments volunteered about the foundry work group. Attention was thus focussed on the relationship, which brought forth more discussion, which again was added to the board in key words.

John		Group
engineer	hostility	older
young	indifference	long service
technically trained	hostility	close knit
		know-how
		pride in workmanship
		rules of membership

In one prolonged heated argument between participants as to whether or not the hosing had been humorous, the instructor apparently literally decided to sit down and let that storm blow over.

Near the end of the period, as the class tried to summarize the situation in which the young assistant must find some way of relating to the group, the instructor raised the question, "Do these sentences apply to Mr. Post?"

At the scheduled working luncheon, another attempt was made to transfer the discussion back to the pedagogical aspects of the morning. Again the participants were so much involved in the administrative situations of the Gordon Foundry and the Dashman Company that little advantage was taken of Professor Lawrence's willingness to discuss how the morning looked to him as a teacher. Something worthwhile was going on and worth cultivating, even though the well-designed plans were being bypassed. Participants were therefore encouraged to study the two cases as well as the related readings for the next day (see Exhibit IV-5).

In many universities these first two cases would have been spaced out over a week, and the students would have had a weekend to ponder over these two dramatic class periods and let them take on perspective. According to Dean Wallace B. Donham there is no way to compensate for this subconscious assimilation over time, "the fireless cooker effect." But instead of a weekend of reflection — even of concern with other things — the Program's 8-week schedule meant that participants went back into the classroom in less than 24 hours.

Next morning opened with the Martin case. It gives information about relations of the General Manager, his Assistant, the Credit Manager, and the General Clerical Section of 25 girls. In 1942 an acute shortage of piece goods was brought about by the advent of war and halted the rapid growth of the company's branded women's wearing apparel business.

After 20 minutes of active discussion, from which the instructor noted brief references on the left side of the board as shown below, he went to the board and suggested, "Let's put down these facts about how the group is behaving." He hesitated a moment before the board considering where to start writing. Then, as the participants recalled various points, he listed them in the middle as follows:

war-uncertainty- frustration Econ. conditions wage freeze Tech) low current value) confused org. structure several bosses no consistent follow-up immediate leadership pressure tactics inconsistency promises unfulfilled less work to be done	lots of breaks lots of talk social events Roth-key person Avoid Byrd Feel justified

The discussion then turned to "lay off some," "let them take time off," "in my experience," at which point, about 50-55 minutes along in the period, the instructor observed, "Now we're coming up with action proposals which will take these things into account" and asked:

"Would it help to feed Mr. Byrd (the general manager) some research evidence?"

This was the first reference to the assigned reading on "maps" and it was made by Professor Lawrence, who observed that "Mr. Byrd has a theory. His experience is a powerful determinant. For him it is reality." Note that it was Professor Lawrence who brought "theory" into the case discussion. Participants who in earlier conferences about the case method had held that men needed "theory" *before* they could discuss a case had plunged into the case exclusively and resisted efforts to consider what might be an applicable conceptual statement.

Argument broke out that Byrd's theory did work under certain conditions. ". . . keep employees in line through their fear of being fired and out of work . . . the problem was insoluble until the war ended . . . At that time . . . get rid of 70% to 80% of the personnel." But he did not shift with changing conditions.

It now being 65 minutes into the period with only 15 minutes left, the instructor went back to the board to check off what

Byrd's theory took into account and then asked attention to other aspects of the situation.

At the instructor's initiative the participants reviewed the case data and volunteered "what we know of output," and "satisfaction." These items he listed on the board to the right of what had been listed about action, behavior of the group.

Productivity
errors increase
low output
meet peak loads

Satisfaction
low turnover
high satisfaction
social

Alternative theories were then advanced, and the period ended in a good-humored, role-playing effort to get Mr. Byrd to accept one of these alternatives that would take more of the facts of the situation into account.

The differing theories were still being discussed over coffee, but more attention was given to Professor Lawrence's report of the problem of getting students to relate readings to cases. Other Harvard Business School Faculty from other courses echoed this difficulty of getting students to shift from thinking about readings to cases and vice versa. Having noted with surprise that these participants, although Ph.D. professors, were not very different, the group went back to the next case to see what it could do in relating the Superior Slate Quarry case to the Research Findings of Professor Zaleznik on *Worker Satisfaction and Development.* (See Exhibit IV-5.) The SUPERIOR SLATE QUARRY COMPANY case is divided into episodes describing conditions at different stages in the introduction of changes in the procedure for splitting and cutting slate.

The instructor asked, "What are some of the things to be considered in determining the productivity and satisfaction of this work group?"

In the first ten minutes many points had been volunteered, and the instructor listed them on the left of the board about as follows:

ethnic mix

community ties

Tech
  physical comfort
  wage payment plan
  2 split — 1 trim
  long standing
    work pattern

economics — other
  products

leadership

After this initial flow of points the instructor looked at the board and inserted *skill,* then asked, "Are we ready for some predictions?"

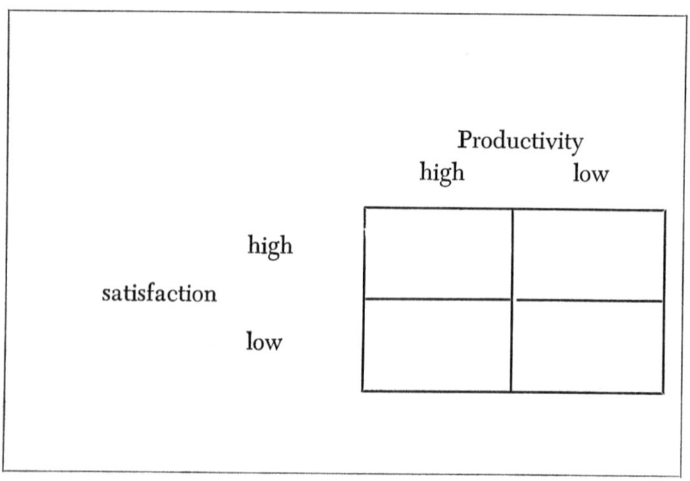

He put on the board the matrix of choices which the class could make as an estimate of the productivity and satisfaction of this work group. Note again that it was the case discussion leader, rather than the visiting professors, who was bringing in the research findings as working tools for analysis.

> There was a considerable amount of questioning and alternatives were suggested to get the projection on a probability basis. Finally, the instructor asked "Could you suggest a better way to get a vote?" "Are you troubled about committing yourself?"
>
> This discussion took about 15 minutes after which a vote was taken, which tallied as follows:

		Productivity	
		high	low
satisfaction	high	4	10
	low	2	4

Then part two of the case was distributed. This described the changes which had been made in the process and layout. After a few minutes had passed, the instructor suggested, "Let us list some of the changes."

On another board he listed on the left, as the group reacted:

> job security?
>
> old teams of splitters
> trimmer removed
>
> output less visible
>
> less skill @ saw
>
> mix unskilled
>
> mill
>
> leadership?

Again, the instructor asked the group to make a prediction as to the mix of Productivity and Satisfaction at the end of from four to six months.

Again there was 10 minutes of sparring discussion about the reality of making such an estimate on so little data. How well can people project how they would like different conditions? What would the leadership be and how might it influence the reaction? But, by the end of 60 minutes a vote had been taken predicting:

		Productivity	
		Increase	Decrease
Satisfaction	Increase	0	0
	Decrease	10	10

The instructor wanted to get the group to consider what we should do with this split in prediction. "Could we start by analyzing the design of the new mill?" Discusion was too widespread and animated for him to channel. Some were pressing for where (if at all) the consideration of worker satisfaction should come into a management decision: At the point of decision to make the investment? At the point of design of a mill? Or only at the point of determining the method of installing the new mill?

Thus, four cases, two readings, but only two days later the participants had become engrossed in the subject, "Management in a Social System." However, the Program was not progressing to the end for which these means had been chosen, namely, the interrelation of "Cases and Concepts" in design of a case-course. During the next three weeks, in which Professor Lawrence continued with 12 more sessions, each followed by pedagogical review, the participants did progress in selecting, analyzing, and arranging cases in their own courses. The men recognized that some transfer had to be made, but they had man's limited capacity for playing different roles at the same time.

Toward the end of this series of 16 paired classes and seminars, therefore, Professor Lawrence took time in class to look at some of the pedagogical issues. He used the cases already discussed to contrast the orientation to action, *what should I do;* and the orientation toward description, *what is involved here.* Both considerations are related in predicting, "If I do this, what is likely to happen?"

The group's experience in predicting the impact of change on productivity and satisfaction led to a more discerning examination of the facts in a case. What is *given* of technology, interaction, present and potential contribution; organization structure, leadership pattern, and backgrounds of individuals — status factors, development, and active need level? What also is *emerging* as group behavior, as a pattern of interactions, activities, and sentiments? What membership roles and norms exist in development, task, social, and security groups?

The research findings and concepts were rarely used as an integrated aspect of case analysis by the participants in their study of cases before class. The period of gestation for such union of concepts with cases probably exceeded even two months, surely more than eight days spread over four weeks. In their final reports, participants expressed appreciation for the opportunity Professor Lawrence gave them to see a course unfold—"the sequence of ideas emerging from the cases." But in another paragraph they recommended: "greater flexibility in scheduling . . . insuffıcent time for reflection and digestion of new ideas . . . many of the morning sessions were too long and the majority of the coffee breaks not truly free . . . a few days allocated for individual case writing, reading, and course development might be considered."

The report went on, "The course development objectives would be better achieved if more free time were given for the individual to work with available Harvard Faculty whose interests most closely paralleled his."

Bringing Together Multiple Objectives

The important thing in beginning to use cases is a willingness on the part of the professor to look on his course as an instrument for active learning—his own as well as his students. This includes a willingness to acknowledge gaps between the concepts, theories, and hypotheses presented in the course and the cases he is introducing. Also needed is the willingness to acknowledge that the subject matter will not provide the entire answer for an administrator at the point of decision; therefore, an instructor also needs a willingness to restate the ideas of his course in such a way as to help the administrator expose the conditions in the situation which are relevant to a decision. In this connection, it is significant that for 40 years, during the development of the case method, most of the cases were written through the partnership of a young graduate who tended to identify himself with the actual administrator from whom he was getting data, and the professor who was going to use the case in his course to help students understand his ideas. Confronted with the data, the professor sometimes had to modify his ideas and sometimes the ideas of the professor helped to give new meaning to the data of the situation.

To repeat, the time of decision brings both the direction of the enterprise and the understanding of a specialized subject into focus. It is unlikely that a professor will find a sequence of actual cases to correspond with his present subject outline. He may therefore have a dual organization of his course for some time, incorporating a sequence of cases with which his students are progressively able to cope and an exposition of his subject which satisfies his sense of professional responsibility. Former participants reported that by trial and error they discovered a sequence of cases in the course which became more and more satisfying because it enabled them as well as the students to learn.

The case provides a meeting point for the administrator taking responsibility for the direction of his enterprise and for the specialist taking responsibility for the design of his subject. In both enterprise and subject there is a system of determinants. There are also points of discretion. These points, or cases, act as the fulcrum on which professor and students can get leverage to work back and forth between abstract subject and concrete action. The attributes to be developed are multiple and compound:

The capacity to generalize — both
policy direction for this enterprise
and
the system of determinants for this subject

Skill — both
for discovery of goals and opportunities
for action in complex situations
and
for logical inference under given conditions

Orientation — both
as responsible leader in action situations
and
as detached searcher for understanding

Maturity (most easily seen in the negative) — both
for control of impulsive reaction
and
for control of obsessive overthinking

Chapter V

Learning with the Case

THIS CHAPTER describes the experiences of visiting professors as they experimented in the classroom and as the group analyzed these experiments; it also includes a limited selection of a wide range of ideas about the learning process and its relationship to administration. As each professor in turn led a discussion, it became apparent that explanations and advice were often conflicting. The evidence was overwhelming that there is no one best classroom style, but that each individual must work through to his own. Thus, the reader cannot be told "how to" lead a case discussion; he can be encouraged to try, and to consider these ideas presented in the Program as he reflects upon his own efforts. Attention to these ideas may help him interpret some of the things which happen in a classroom as he tries different approaches in case discussion, but to focus on teaching alone is too narrow a concept. Between the first and the later experiences in leading a case discussion, participants showed marked improvement. Partly it was a matter of practice, but of more importance was their involvement in the whole case method complex. The professor's own demand for internal consistency became a powerful stimulus for learning and for integration of his own pattern of activities in the study of business administration.

Thus, although most inquiries about the case method seem to come from curiosity about what goes on in the classroom, this book has delayed discussion of the case in the classroom until other elements in the set of related activities have been clarified. In the Program, the reason for classroom action could be seen more clearly by reviewing how the case researcher became aware of the way his interests and behavior influenced the data he got in the field situation. Then the writing of the case also made each participant keenly aware of the way his own opinions had become mixed with the data, and how difficult it was to edit out evaluative slants. Finally, the finished case made the professor reconsider his course objectives and conceptual scheme. So, by the time a participant came to lead the discussion of a second case in class, he was aware that somehow his interests and behavior were going to be involved with that of his students in a complex process.

Much has already been written on leading the discussion of a case.[1] Furthermore, many of the 223 published case books have excellent suggestions about the use of cases in the books themselves or in manuals for instruction.

Seemingly, many of those who inquire about "how to teach cases" think they will be teaching by the case method as soon as they master a technique of discussing a certain kind of written material; the case discussion is assumed to be the epitome, if not the whole, of the case method. On the contrary, the Program found that increased effectiveness in using cases came from better understanding of the total case method pattern. Understanding and proficiency developed together. For this reason, participants in the Program were plunged into all activities of the case method complex simultaneously.

However, the classroom is more than, and different from, the re-creation of the professor's experience of getting the case data in the field. The class room brings together a large number of

[1] *The Case Method of Teaching Human Relations and Administration*, Kenneth R. Andrews, Editor (Cambridge: Harvard University Press, 1953); and *The Case Method at the Harvard Business School*, by Malcolm P. McNair and Anita C. Hersum (New York: McGraw-Hill Book Company, Inc., 1954).

individuals of widely different backgrounds. Their relationship to the professor, individually and as a group, and their reactions to one another must be considered. Thus, in this chapter, we are adding a new element to our case method model: the student. To the study of administration in a concrete living enterprise, and to the design of a course relating specialized knowledge, we must now add a compatible view of how students learn to take responsibility in this complex process.

The present chapter consists of five major sections. The first is the analysis of "The Teaching-Learning Process" presented to the Program by Dr. Nathaniel Cantor. Then a report on the actual experience of the participants in class leadership — an experience which was concurrent with field research, case writing and course design. The third section of this chapter deals with "sequential considerations" before, during, and after a class session. Fourth are some significant analyses of the levels of class discussion made by Associate Dean George F. F. Lombard in a series of talks to the Program. The final section concerns "the use of differences" in teaching-learning by the case method.

"Learning Which Makes a Difference"

Participants came to the Case Method Program with preconceptions which created a dilemma about behavior in the classroom. "Should I be directive? Or should I be permissive?" This is indeed an awkward axis of controversy and a difficult one to avoid. Fortunately for the Visiting Professors Case Method Program, one of the coaches, Pearson Hunt, suggested that Dr. Cantor be invited to talk to the first group of participants on August 4, 1955. He came back each year until his untimely death. Somehow in his very earthy manner and speech, Dr. Cantor took the group away from the fruitless question of whether to be permissive or directive. He was able to focus on the emotional context in which significant learning takes place, the responsibility of the instructor for creating this atmosphere, and the importance of this context for the student's assimilation of intellectual content.

After ten years, the tape recording of Nathaniel Cantor's memorable first speech emphasizes how much of the impact was in his own acceptance of his precept that the emotional context is what makes learning significant. Because his talk was such a vital experience for scores of participants, excerpts are presented from the verbatim transcript (which he gave permission to use), even though they lack in typed form the intensity of the direct interpersonal communication. The title used for Dr. Cantor's talk was "Learning Which Makes a Difference."

> . . . I want to take you now into a quite different dimension . . . the teaching-learning process of which the instructor is the final differential factor. . . .
>
> The teaching-learning process . . . for purposes of analysis here, may be distinguished by three various aspects. What is happening to the student? What is happening to the instructor? And what happens between them? That is the teaching-learning process. . . .
>
> For example, what is happening to you now in relation to me and to each other? What are my feelings and there are plenty, with regard to you? And what is taking place and will take place between us? Now, when you analyze that out, including process and content, you will have an awareness and an articulation of the teaching-learning process which has rarely been done. . . . A kind of analysis which would require two, three, four, thousand pages.
>
> But I am going to try and limit . . . myself this afternoon to what are the characteristics of the skilled teacher. . . . (You can't separate these things easily.) . . . Another way of saying, what must the teacher be aware of in the classroom or in the teaching-learning situation? . . . May God forgive me for what I am going to try to do in the next five minutes. . . . These ideas . . . are common property of a great many people. . . .
>
> Now one of the greatest confusions . . . is a failure to distinguish between knowledge and understanding. If I said to you, I'm going to give you knowledge. Now I say to you, oogle is oop. This of course is called Oogology. . . . And later on in the week I give you a piece of paper and I say to you, will you please tell me what oogle is? No copying please. And you tell me . . . oop . . . and I congratulate you. You got 100%. . . .
>
> I maintain that genuine education is a matter of understanding, which is not even primarily a matter of logic. It is not the transmission of facts from one to the other which constitutes education. It is translation of the data in the palpitated living realities of

the learner's life, not the transmission of facts. Because you never get a fact unless you, the learner, assimilate that fact to some outgoing need or problem which is your own. . . .

Understanding must stem from the frame of reference of the learning. . . . Every fact must have a frame of reference. I say to you, what are the facts in this room? That question is unintelligible as stated. . . . From what point of view, from whose point of view? You'll say, the facts in this room are desk, table, people. I would say, the gravitational influence of the sun or the bacterial action in my lower intestines now or television rays. Well, of all facts why do I select bacteria in my colon? Because I happened to read an article last night which dealt with this problem and I am now interested. You will select data which allegedly answer a problem, however articulated, which you have. And you don't have understanding unless you take that knowledge and assimilate it to an entire organic activity. . . .

Another premise is, you never learn anything unless you are disturbed. Our friend Newton has a law of motion, a body at rest remains at rest until disturbed by some outside greater force. . . . It is the law of inertia. . . . Another way of saying this is, every system in nature tends to remain in equilibrium. That is the law of all forms of nature, of education. Here is a piece of paper. It is relatively at rest. I want to disturb it. I do disturb it. It goes back into equilibrium. I'll try again. It goes back into equilibrium. Now a little plane geometry, a little history, a little marketing and the student remains pretty much the same. . . . Essentially the little organisms tend to remain constant. . . . *To learn, means to change.* If I were to change this paper's organization I have to, by definition, change its structure. Now observe. [Crushing the paper.] What you are hearing on one dimension is the breaking up of the crystal structure of this paper. It makes a new structure at the expense of a great deal of disturbance. Of course you can't do it with the students. They think you are psychotic. So we have to do it a little more gingerly. But if you are going to change, well that's what it costs if you are going to learn. . . .

You will never learn without confusion and disturbance. And the reason we find it very difficult to learn is every one of us wants to resist changing, because if you change you get confused. Your system of protective devices, your defenses, your whole security system of self-esteem, all are pretty good. . . . I saw in a class a group of learners, and this is the way some of them were learning. [Lounging back in his chair.] Me? Not me. Nothing is going to happen to me. Or . . . if a girl says to me, do you want to kiss me? And I said, well — sure I do. I think that she ought to doubt my intentions. I'm not very much interested in kissing her if I am sitting this way. But if I do want to kiss her, I think that I become active, I go forward. The kind of physical mode of attention you observe in any class is a pretty good indication of one's willingness to learn. . . .

We become active if we want to change.

We want to change and we don't want to change. . . . This is the constant endeavor of one to have his cake and eat it too. We are built that way. We don't want to be educated, and we take the easiest way out. We talk, the most wonderful device of escaping change that I know of, using language to prove anything, to disprove anything, to talk your way in or out of anything. We have lost our tails and we wag our tongues. . . . Talking, talking, talking. The language is the finest weapon we have of not communicating. It might lead to communication but that is another story. So you do not learn unless you are disturbed. . . . In that sense I mean all learning is in emotional context. . . . Learning is personal. . . . I don't think anybody can learn except when he is alone.

There is plenty of argument and plenty of talk but I think, frankly, learning makes a difference. You notice how I am emphasizing that. You see, I don't mean memorizing is learning. It is a kind of learning I suppose. But I mean significant learning which makes a difference in behavior. . . . You go back and forth and you discover what was said many years ago, wisdom is acquired through affliction's school. You get wise through . . . affliction, through discipline's suffering, through being bothered, irritated, and annoyed. And using that confusion and annoyance and irritation . . . and you do that with lead in your belly and hot tears down your cheeks when you are learning. But I can teach you oogle oop without too much disturbance, or how to market. . . . If, for example we should . . . analyze out the underlying, hidden, ethical, moral, political and economic assumptions behind the work you are doing here, the ethical implications of the whole case method approach or any approach of the schools of business administration and make articulate what premises underlie that, believe you me you'd be a fairly uncomfortable group of men. . . . a genuine assimilation of respect for people in industry has got to do something radical to our modern industrial life. . . .

Learning starts where the learner is, not where the instructor thinks he should be. . . . No one learns from someone else's experiences. It is even questionable whether you learn from your own. But you certainly don't learn from anybody else's experience. And you don't learn from your own experience unless you go through the experience to learn. And you don't learn by doing. You become a monkey by doing. You learn by doing what interests you. . . . A frequent mistake I think we have made in education is not to stop where the learner is. The moment you use a textbook you are saying to the student, be where the author is or be where *I* say the author is. So you sit in your room and you talk about the archaeological remnants of antiquity. . . . The statues of Egyptians have little to compete with Florence because Joe is next to Florence. He is not in antiquity. . . . Now we are doing that a good deal of the time. . . . It would be very interesting to make sure when anything is being discussed that you stop where the student is. . . . To give you an example, of one of my own classes I have had in Crime, Crime and Society. So I come in and I say, how do you do? And I sit back. The students that never had me before start looking around. . . . Then some great soul says, well, isn't this the class in Crime and Society? I say, yes. Well, why don't we start. I guess we are going to discuss Crime and Society. Well, why don't we? I say, why don't we. Well, aren't you going to tell us? Tell you what? About Crime. What do you want to know about crime? They always say, what kind of a screwball have we got here? What happens then is somebody will say, why don't you give us a few definitions? I say, what do you want a definition for? Well, to start. Oh, you want a definition to start? And I go to the board and what, what is etiology of definition? If I give you a definition we will be trapped. There is nothing to talk about if you have a definition. You know I don't give a definition. The reason we are here is to find how to define crime about one year from now. I do this a little more slowly. Then someone says, well, let's talk about it. What is a crime? Well when you do something bad. Like taking off my trousers? Well, yes. Taking off my shirt? No. So if I run down the street without my trousers, would that be criminal? I think so. Why? . . . I am going to try and be where the students are and not too far ahead of them. Always you start with what they know and feel and move forward with them. So much for the teaching-learning process.

Dr. Cantor challenged the group each time he met with them, and afterward, his work continued to help professors visualize how the objectives of their courses might actually be furthered by shifting from "teaching" to fostering the conditions for "learning." Some detailed illustrations already have been reported in Chapter IV. But Dr. Cantor pressed on to make the professor aware of his involvement in the process as a person. With the passage of years, the single most memorable thing about the Program for most alumni was the way in which this role of the professor as a person had been illuminated.

I certainly think the skilled teacher should always guard himself against using the students for his own needs to control and dominate. The skilled instructor tries to keep the learner's feelings at the center of learning. . . . Now mind you, I'm not underestimating content of education. . . . Knowledge is too important to be passed by as a bloodless concept, as a verbal systematic relation. It is the import of the concept to the student. And by keeping at the center how the student is feeling, how he is reacting to what is going on intellectually, trying to capture what the thing means to him you are assured, it seems to me that a more favorable act . . . will occur. The skilled teacher fortunately tries never to make moral judgments about his students, but he seeks to understand. . . . You hold a student responsible, yes. . . . A skilled teacher recognizes that all significant learning can only come from the creative efforts of the learner. That's another way of saying learning is personal. You cannot learn for anybody. . . . We turn it around. We ask the students to tell us what we know. We know the answers. . . . That's no education. It's really amazing when you start to think about it. . . . Essentially the student must be the one to raise the significant problem for you to help him find the answer . . . what I want to convey by *skilled teachers* . . . is being different. I suppose that psychologically the greatest and most difficult problem in human nature is to abide difference, to be identified with people on one level with the identification so fine that you can abide all kinds of differences. Now if you are indifferent you don't care. . . . You have to care for the student, which brings you to this last thing.

Unless the teacher respects himself, he doesn't respect the student. . . . A person that respects himself as a living creature, having struggled to obtain something, will feel something about himself. He'll stand for something real in his own life, not something fake, something real, that's got to go out and communicate itself. You never fool a student. You never fool each other . . . emotion always responds to emotion. . . . By your day to day detailed relationships with people, they feel your respect or they feel your disrespect. You create atmosphere when you say to your students . . . How can you use me to help you? And you have a right to fail. You have a right to be different. You have the right to grow in your own way.[2] My job is to free you, to help you get rid of the anxieties, the vicious comparisons, the fear of being ridiculed, to free you to dare to be yourself in relation to another helper.

For those who would like to see a more systematic statement of Nathaniel Cantor's ideas, it might be of interest to note that Carl Rogers[3] refers to Nathaniel Cantor's book *The Dynamics of Learning* (Buffalo: Foster & Stewart, 1946). He cites Cantor as one of the pioneers in the application of modern psychotherapy's research to the tasks of education. In the years after his death, Cantor's estate gave permission to reproduce and distribute to participants Chapter 12 from his book, *The Teaching-Learning Process* (New York: The Dryden Press, 1955) which included "The Propositions of Modern Learning":

(1) The pupil learns only what he is interested in learning.
(2) It is important that the pupil share in the development and management of the curriculum.
(3) Learning is integral.
(4) Learning depends upon wanting to learn.
(5) An individual learns best when he is free to create his own responses in a situation.
(6) Learning depends upon not knowing the answers.
(7) Every pupil learns in his own way.
(8) Learning is largely an emotional experience.
(9) To learn is to change.

Dr. Cantor's analysis of learning "which makes a difference" is especially challenging to students of business administration, soon to be in a position where their willingness and capacity to take responsibility for decisive action will be essential. Therefore participants, after hearing Dr. Cantor, more eagerly explored the concrete manifestations of this point of view in the classroom and class period.

---

[2] Dr. Melvin T. Copeland who joined the Harvard Business School Faculty in 1910, in reading this manuscript observed that President Lowell of Harvard stressed that, "The only real education is self-education."

[3] *Client-Centered Therapy,* by Carl R. Rogers (Boston: Houghton Mifflin Company, 1951), in Chapter 9, "Student-Centered Teaching," pp. 384–428.

Trial, Observation, and Feedback

The basic approach was to give each participant in turn an opportunity to lead the discussion of a case, during which the other participants acted as students. The procedure was to follow the class discussion with a pedagogical conference on what had taken place. Thus each summer there were from 20 to 60 analyses of immediate teaching-learning experiments. These episodes always aroused interest and the desire for more wide-ranging commentaries on the whole learning process. The professor obviously does something different in getting ready to lead and discuss the written case in class from what he does in preparing to give a lecture and even different from getting ready to lead the discussion of a subject. The Program gave each man freedom in class preparation, but at the same time made available whatever advice he might want from his coach.

The case, the professor's objectives for the course and for a given period, the preparation made by participants for the case discussion, the dynamics of the professor-student group in class and the cumulative atmosphere in the institution — all these variables interact. Their very "togetherness," which makes the process so baffling, is indeed the essence of administration for which professors are preparing their students. Although isolation of one attribute may provide a gratifying neatness, it is mastery of the whole that is being sought, and the whole constitutes something different from the parts.

The task in the Program was to get participants immersed in the total case method pattern and still find some way of feeding back to the individual a more useful understanding of where and when in this system he might make a difference. Each morning one or two participants took a turn leading a case discussion. These experiences provided repeated opportunity to see the many points at which the professor had a choice. But the participants who were acting as students could see this much more clearly than the one who was playing the instructor's role.

Feedback

One of the unique features of the Program was the possibility of the instructor's learning directly from his class how it looked from the "learner's" point of view. How much the individual instructors really wanted to know about the reactions of their students differed considerably from one participant to another.

Perhaps the complexity and difficulty of observing and interpreting what went on in a class could be suggested by referring to the videotaped closed-circuit television playback of case discussions in 1964. This seemed, at last, the answer to Robert Burns' wish: "To see ourselves as others see us!" Without detouring to discuss this experience, it is important at least to note that during the viewback of a case discussion when one participant said to another: "There! See what I mean when I say you are a Cheshire cat?", the group became more self-conscious than was anticipated or could be constructively managed in the Program as it was designed.

Less sharp tools of observation seemed to be more tolerable and allowed the life of the group to mature in a different way. The stress in a classroom under ordinary academic protocol might be just as great as during the case method program, but there participants rarely have an opportunity to know what these stresses are. Even without the vividness of the videotape, groups in the Program became critically involved enough that professor-participants reached the point in one case discussion when a man who could not wait to be recognized provoked the speaker to shout: "Shut up! Don't you ever interrupt me again." And countered with, "Well, why don't you say something?"

Only once did a professor actually get up and walk out of the room. More often a man might silently withdraw into himself and let the world go by. These melodramatic incidents, in the minds of participants, were often counterbalanced by what men described as deadly dull, repetitious belaboring of the obvious.

The role as hosts to the participants always left the coaches in an ambiguous position. Participants professed a desire for more feedback, and coaches always kept their doors open for consultation. Experience in the classroom showed a large, emotionally loaded, subjective component and even observing the process was far more complicated than participants were normally prepared to acknowledge. As leader or as student, each man wanted to know what the others thought of him, yet at the same time he did not want to know. To minimize the question of who should take the initiative, strategy meetings between coaches and teams were tried in neutral conference rooms.

In various years the schedule included a period after the case discussion to review what had been going on. During the case discussion all participants tried to concentrate on the administrator's responsibility in a given situation and what it would be useful for him to do, or what one of the participants would do if he were in that position. In the review session an effort was made to take a look at the pedagogical significance of what had been going on a few minutes earlier. Sometimes the coach of the participant-professor leading the case discussion would preside at the review. Somehow the term "critique session" came into use in spite of every effort to shift the emphasis from criticism of the instructor to analysis of the pedagogical process.

The weekly schedule referred to these review sessions as "criteria sessions." The ground rules were that the inevitable spontaneous criticism could be voiced if the speaker identified the criteria by which his likes or dislikes were prompted. This effort to expose fully a man's criteria for critical evaluation required more concentrated discipline than the participants were ready to accept.

A partial explanation for the practice of having the discussion leader continue to preside at the review session may be found in an incident which occurred early in the program. One genial participant pushed aside his coach at the beginning of the review period with the comment, "I would rather be talked *to* critically than to sit here and be *talked* about." This point of view had an almost undeniable force, even though theoretically all might agree that the group was reviewing not only the conduct of the leader of the case discussion but also the part played by each participant. Maybe there was a self-protective recognition that no one had been doing his homework in preparation for the case discussion as one would have the right to expect his students to do!

In addition to the review-criteria-critique sessions, an audio-tape recording of the class session was generally available for playback in the audio-visual laboratory. Sometimes the leader would invite his coach to listen again to the session, or would listen alone, or with one or two friends. As a guide to anyone who might want to do further work, one might note that traumatic as was the "cheshire cat" episode, nearly everyone wanted to see the playback of his videotape, whereas very few participants wished to listen to a playback of the audio tape.

Classroom Observation and Process Analysis

Observation of the classes in the Middle Management Program was a major feature of the first few programs. Prior to arrival of the visiting professors, the Harvard Faculty teaching in the Middle Management Program had discussed this possibility with the MMP students. These men had been on the campus for five months and were therefore already well acquainted with each other and with the Faculty members. As experienced business managers, they were not concerned about having visitors. They and the Faculty agreed to welcome all 18 of the visiting professors at one time if the professors wanted to attend as a group. But for most of the years in which the case program overlapped with the Middle Management Program, only four or five visiting professors were scheduled to visit a class at one time.

The Faculty who were teaching in the Middle Management Program met periodically with the visiting professors as a group to discuss their teaching objectives and plans. Subsequently they came back for further discussion with the participants who had observed their class sessions. Their genuine interest in feedback from the visiting professors helped to set a climate in the early years in which case discussions of the visiting professors also could be reviewed constructively.

In each succeeding program it became more evident that the participants actually were much less interested in observing Harvard instructors than they were in an opportunity to participate actively themselves. Thus when the Middle Management Program terminated, there were still ample opportunities for observation for those who chose to take advantage of them. This opportunity was thanks to the willingness of Harvard instructors to admit occasional visitors to the variety of short management programs being conducted on the campus. The Industrial Research Institute and the International Marketing Institute were especially cooperative in welcoming observers. Coffee periods together allowed the participants of the case method program and these management programs to interpret for each other the perplexities and values of their experience. Other Harvard Faculty members generously met with the visiting professors to demonstrate their approach to classroom conduct by leading case discussions with the participants and then staying on to discuss their reactions frankly. In addition, Richard B. Donham, then Dean of Northwestern University School of Business Administration, kindly led the discussion of a case early in one program as a way of evoking concrete questions about the case method.

Although video playbacks of the visiting professor's own performance were difficult to handle, the few videotapes that were available, of actual case discussions in other groups, were useful. These taped "classes" could be stopped at will to consider what was going on and the alternatives that were then open to the instructor or the participants.

Efforts were made in several different Programs to bring some system of "process analysis" before participants for discussion. Frequently coaches who had a personal professional interest in such group relationships commented during the criteria period with helpful remarks about what had been going on within the group while a case was being discussed. In the first Program, in 1955, Professor Robert F. Bales, Director of the Social Science Laboratory at Harvard, talked with the visiting professors about his experience in observing groups. In subsequent years copies of his article,[4] on conference leadership, participation, and observation, were distributed. On occasion a Doctoral Candidate observer made a more systematic time analysis of interactions initiated by the instructor and by the participant-student for subsequent discussion by the group.

Participants knew that significant learning was going on in

[4] "In Conference," *Harvard Business Review*, March-April 1954, pp. 44–50.

the class discussion of a case but the process was elusive. Mere classroom observation failed to involve participants enough to experience it. On the other hand, it became so dominant when men were actively participating in case discussion that they could not see their own part in the process. However, abstract analyses of the learning process were difficult to relate to the shifting roles of participants in an active discussion. Some experiments at feedback have been briefly reported. Further experimentation and research may clarify the complex phenomena of the classroom and give the instructor more insight and help him control his role in the learning process. Meanwhile, a few more suggestions from the Program may be useful.

Sequential Considerations: Before, During, and After Class

The repeated experiments of scores of visiting professors revealed that there was much room for innovation and individuality. These experiments also showed that choices had consequences with which the instructor then had to live. Even the most obvious of the turning points suggest the opportunities for variety in classroom conduct, and some possibilities for control by foreseeing additional points of choice. These points of choice are almost too obvious to mention except that from repeated observation their interrelationship to each other and to the multiple and compound objectives of business administration were often overlooked. Consider then the "obvious" choices the instructor can make to influence learning before class, during class, and after class.

Preparation for Classroom Discussion of Cases

Dr. Cantor implied that professors need to shift away from "teaching a case." Students also need to shift away from "learning a case." Nevertheless, there is need for rigorous preparation, both for the professor and for the student. Even with a case that is comprehensible to the students, it is still a major task to help them see what is involved in being well prepared to discuss it.

Over the years, as experienced case teachers served as coaches in the program, they came to agree that the Visiting Professors were under a handicap because the participants, when acting as students, frequently were not well prepared. This was not because they lacked interest or effort but because many of them had little idea of what it meant to prepare for the discussion of a case. In the Harvard MBA program, effectiveness in the classroom has always been enhanced by the cumulative impact of five or six Faculty members consistently communicating to the same students in their classes what it means to be well prepared for the discussion of from 10 to 15 cases a week. Thus, even early in the first term of the MBA program, students would have had far more exposure to case preparation than the participants of the Case Method Program at the end of eight weeks.

For the 1964 summer Program, Professor Milton Brown was persuaded to start the Program by dealing with the preparation necessary to discuss a case. Professor Brown had been a highly successful coach in two previous summers and realized the delicate line between acting as host and acting as instructor, particularly early in the period when visiting professors were getting acquainted with each other and with the Harvard coaches. The sessions with Professor Brown on case preparation were announced in the schedule as follows:

Monday, June 15

9:00 a.m.	Aldrich 209 -- Orientation with emphasis on case preparation. Briefing by Professor Milton Brown for preparation of today's case. (To be distributed by him during the briefing.)
10:30 a.m.	Individual preparation for discussion of the case.
1:15 p.m.	Small group conferences on the case.
2:15 p.m.	Aldrich 209 -- Discussion of the case led by Professor Milton Brown.
3:30 p.m.	Aldrich 209 -- Discussion of the discussion with particular emphasis on what it means to prepare to discuss a case.

Professor Brown selected for discussion the Fidelity Mills, Inc., case. It had been written for the Advertising course and reviewed the entire marketing program for a new dish towel during the period 1957–1961. Sales had increased steadily but still were well below original expectations at the time it was introduced as part of the product line in 1957. The administration had to decide whether to continue the product and, if so, how to increase sales.

In a brief introductory comment about preparing to discuss the case, Professor Brown said he found it useful to think of a case as a statement of relevant facts. The task of the student was to define the problems faced by the company and to consider alternative solutions. The instructor would expect the students to have arrived at a decision and to have thought through an action procedure.

Professor Brown pointed out that the student-participants should not be reading the case to learn the facts for their own sake. They should read the case with *use* of the facts in mind. As the instructor, he was going to be interested in the reasons why his students thought a fact was relevant. Therefore, in preparing to discuss the case, he asked students to be "formulating your own statement of the problem."

Professor Brown had come to this session directly from a meeting with a group of business executives who were at the beginning of an executive development program. He reported how he had told them that they should be prepared to discuss the case just as if they were in a business meeting. There would be a difference, of course, because in the time available it would be unlikely that the class would come out with one solution, but he would expect each man to make his own decision. "You won't have all the facts you need, but we are going to be *working forward to action* in this situation rather than looking backward from results or somebody else's decision."

After this preview and about three hours for thorough preparation, Professor Brown met with the group in the classroom. At the appointed time, he asked: "Where do you want to open?" Professor Brown was, in effect, taking Dr. Cantor's approach of starting where the students were. After a few moments of hemming and hawing by the group, someone volunteered with the statement: "I'll take a leaf from your book. Let us start with the question: 'What is the problem?' "

As statements were offered, Professor Brown scratched on the board only a key word or two and interrupted frequently to ask "What evidence?" or "Why?" For instance, after 10 or 12 minutes, one of the participants pointed out critically that the target percentage of profit stood in the way. Professor Brown pressed him, "Is there any evidence?" A flurry of hands around the room was apparently ignored as Professor Brown followed this man's line of thought that "a smaller percentage of the market might lead to other channels of distribution." It appeared that Professor Brown was much more willing to give this man time to think out loud than were the rest of the class, who were visibly impatient at not being given a chance to talk. Paradoxically he was permitting expression by being directive.

After about half an hour, in which a number of other participants had had the opportunity to carry a line of thought as far as they could under Professor Brown's questioning, someone broke in with a declaration: "We are only dealing with symptoms." At this point Professor Brown walked across the room toward him and encouraged the man to go on. After a few minutes, Professor Brown said something to the effect: "Yes, I hear what you're saying, but what do you *do* with this towel *now*?"

In the late afternoon session after shifting from the business conference atmosphere to the discussion of pedagogy, Professor Brown said he did not really know what the visiting professor had meant at that point and, therefore, had fallen back on the action question. This brief episode is significant, and typical of the way in which the need for action in the case provides a common reference point to which the instructor and the class can return when necessary to get on the same wave length. Its existence provides a common reference point. In this episode, for instance, the discussion went on to the question of lowering the price, which Professor Brown followed up with a line of questions beginning, "If you lower the price, are you willing to expand the plant?" The class was brought back to additional evidence in the case as an actual enterprise rather than as an abstract statement of a hypothetical question.

The case is distributed in advance *not* so that students can *learn* the case facts. The case facts are like a catalyst, essential to the process but not its purpose. The case is distributed in advance so that all students will have the same concrete system before them and become aware of the meaning to them of the things, people, and events in the case. Discussion of case after case cumulatively during the course helps the students get more significance from the facts in each case. In the process of case analysis the student acquires the concepts in a course, and sees the need for a more precise knowledge of the subject. Knowledge of this subject in general, however, does not take the place of intensive examination of the facts and their interrelationship.

Chapter III described the difficulties professors have in leaving opinions *out* of the written case. Similarly, professors, as students, had difficulty in dealing with a set of facts to which no authoritative opinion was attached. Although Professor Brown in the morning had said that the facts were not to be learned as a text, they were to be *used,* participants had difficulty in seeing this difference. Professor Brown had so assimilated the multiple objectives and attributes of administration that he turned from one to another as opportunity arose in the discussion. Such random access to course concepts was a novel shift from the usual sequential programming of a course.

Look, for instance, at his response to the participant who was giving a little lecture on pricing theory. Rather than argue the theory, Professor Brown simply asked him, "What would you do now — in this situation?" Professor Brown was asking this man to take the administrative point of view and show the relevance of his theory for action. At another point, when a participant was trying to relate the facts in the case pertaining to the desired percent of the market and to the channels of distribution and the target profit percentage on sales, Professor Brown held off the rest of the class to encourage this effort at situational thinking.

These episodes illustrate an important feature of the case complex — the use of theoretical marketing models in relation to a specific case. Professor Brown brought the consideration of marketing as a subject together with policy direction for this particular company as a total living enterprise. Thus, the bare facts from the field situation began to be reconstituted in the classroom. Because this was the first case studied by the group, Professor Brown pointed out that he did not expect the class as a whole to come to a decision, but he did expect each individual to make his own decision. Thus, in class he made only enough of a scratch on the blackboard, as he said, to "remind myself of what had been brought up."

The purpose of the sessions with Professor Brown had been to help professors new to the use of cases think about preparation for class. Professor Brown was demonstrating, by the way he wanted "students" to reason with facts of the case in class, what they needed to do in studying the case *before* class. Presumably if these students were to meet with Professor Brown again they would have had one day's experience in class to guide them in getting ready for the next class. They could see that they needed to have the facts in mind and how the facts were related to each other and to the direction of future action.

Choices Facing the Instructor at the Beginning of Class

The obvious need to open a case discussion provides one point at which various styles may be observed. The Program was not a research project for which data were accumulated, but from time to time notes were kept. An experiment one year with course development led to a set of fairly careful notes on how different men began and acted at other points in the discussions for which they were responsible.

That summer the 20 participants were divided into 4 subgroups, each of which took responsibility for a short course somewhat as described in Chapter IV. Noting the choice which the instructor made to open the case discussion provided an opportunity in the feedback session to consider the relationship between a man's intent and what he actually did. The case method does not have one model but a complex and flexible system of choices from which a professor creates his own pattern. In this crude classification of opening choices into five "styles," the constructive intent is obvious. The possibility of negative interpretation by the class is noted. The purpose is to illustrate the range of choices used by a number of instructors experimenting with the use of cases. Most experienced instructors will use all five and more styles on occasion.

The first style may be identified by those visiting professors who seemed to open the class meeting by directing attention *away* from the case, to the subject. This type of opening failed to reward the student who had spent time the night before analyzing the case. The opening words and actions are shown on the left. Noted to the right is the use the instructor made of the blackboard.

"What were the objectives of the case writer?"	Little or no use of board during class.
A 7 minute explanation of outline, "Any comments?" Then, "Who are these agents?"	Before class, on side board had summary of previous cases in course.
	Before class on main board put three questions: (1) What are characteristics of market? (2) What is company looking for in its channels? (3) What channels best meet these requirements? During class: lists 20 characteristics.
"What were the objectives of the case writer?"	Before class: summary of inventory control process.
"Let's start by pinning down our concept of control."	Arranges on board from Socratic dialogue the elements in cycle of control and concentric records.
A 20 minute review of case, theory, and technical notes. Then, "What policies governed their operations, costing, and pricing?"	Before class has reproduced table on another board, "What role do costs play in pricing?"

Another style was demonstrated by professors who went to the other extreme and started off with the action question: "What should be done?" Under the mores of the group, the visiting professor who had spent hours the night before preparing thereupon gave a lecture for the next 20 minutes or half-hour, while the rest of the class became restless at having no way to show their agreement or disagreement. So while one "good student" was being rewarded, the instructor was losing the rest of the class.

"What number should Shay tell Thompson to make?"	Arranged elements coming from discussion of technical process in relation to decision model.
"How many should be made?"	Lists comments from class arranged by steps in production process.
"What decision have you made?"	Lists from class questions which must be decided.
"What decisions have you made?" "What questions will company have to decide?"	Lists questions and identifies them with appropriate labels from text.
"What was the proposal?" "What should be included?"	Three boards used *before* class to note conditions and agenda for class.

In his talk from which excerpts appear later in this chapter, Associate Dean George F. F. Lombard observed from his experience with this phenomenon that the "normative" question "What should be done?" seemed easier to handle later in the period, after the class had obtained a deeper appreciation of what other students thought had been going on in the company.

A third category of classroom openings apparently stemmed from some realization that there needed to be a common understanding of the facts. The professors who felt this way gave a 10-minute to 15-minute explanation of the facts in the case. This killed the interest of those students who were well prepared to discuss the case, and confirmed the impression of the others that it really was unnecessary to do much home work for this instructor.

A 10-minute review of facts from board. Then, "What is the problem?"	Listed on board key facts *before* class.
A 14-minute talk about "givens" on board. Then, "Now we must decide on__________"	Before class had "givens" on one board; previous cases on another; and chronology and questions on another.
A 10-minute explanation of critical implications. Then, "Look at page 7, is this a fair statement of the basic problem?"	During class lists 9 considerations or conditions determining action on "the basic problem."
A 10-minute explanation of fixed and variable costs and their effect on business decisions. Then (1) "Did he make a correct analysis?" (2) "Are his 'instructions' adequate?"	None

A fourth group of openings were those used by professors who sought order by imposing on the discussion of the case a "problem solution model."

"Mr.__________will you please contribute issues you see in the case and then we will pick the most important one."	Lists issues, but has difficulty in shifting from listing to evaluating.
"What is basic problem?" (At end of class, "How do you vote on alternative?")	Lists comments from class under: Problems Alternatives
"What are issues?" (At end of class, asks for vote.)	Lists: Issues; Decisions
"What do you think are the prime issues?"	Organizes comments from class into areas on the board which came to be identified with major issues and alternative courses of action.

Typically cases for the Program were selected because the

purpose, the direction, and the major values in the situation were *not* known. The attempt to hold the class to the instructor's predetermined criteria or "givens" proved frustrating both to the professor and to the participants and led to many different problem-solving discussions going on concurrently, to the bewilderment of the instructor. Dean Lombard's talk, which follows, also dealt with the elemental step of inquiry, which typically helped a case discussion organize toward a more manageable progression.

A fifth style of initial actions by the instructor was characterized by a more open invitation to students to discuss their impression of what had been going on in the company.

What do you think of the controller's report?	Listed 5 ways in which the report might have influence (part of only one board).
"What is your general reaction to this case?"	Little or no use of board during class.
"Who is Ralston Purina?"	"Would it be helpful if we put organization chart on board?" At close of class adds Sales Manager to organization chart.
"From previous cases we saw problems of coverage, cost, and control of sales representatives, has this sales manager made any progress on understanding these problems?"	On board listed class comments under three headings: Objectives Policy Concrete Actions
Made short summary of cases thus far "to identify approaches and problems through critical factors," then, "Mr.__________, please give background facts and issues."	During class listed class comments made under five headings: On main board under: Facts Issue Decision or Opinion On one side board: Broad General Issues On other side board: General Knowledge
"Is there a problem?"	Little or no use of board during class.

Professor Brown's discussion of the case, as reported earlier, demonstrated that this open approach might require agility and use of the blackboard by the instructor in order to help organize contributions from different men, but would quickly involve the class as a whole. Also, in order to encourage advance preparation, the instructor might ask a student to relate his contributions to action that he would take.

These openings surely do not exhaust the possibilities, and some variety from day to day adds to class interest. But perhaps this partial listing of options is sufficient to arouse curiosity about how a professor may find and encourage order in the discussion of a case, even though the dynamics of the class are quite different from those involved in a logical lecture. For instance, one of the first shocks coming to an instructor using the case method is to find his students less well prepared than he had assumed. Some choices of approach to the class make this discovery less an obstacle than others. Another perennial choice the instructor must make is whether to provide background materials by reading or lecture before or after cases involving such knowledge. The probable consequence of asking the students to study a case and then, instead of putting them to work in class, requiring them to listen to his lecture, would be a decline in preparation.

The relationship between student and professor in development of a sequence of cases is a basic and inescapable reference for all these choices. What happens to the relationship between professor and students when students are asked to take alternative sequences of activities before class, during class, and after class? The feedback period in the Program helped instructors see how their "helpful" lecture undercut the homework their conscientious students had done. It was clear that less time would be spent in "working for that so-and-so" next time.

Choices at the Point of Repetition

Almost any way the case discussion starts, student participation will spread out until a wide range of considerations have been covered. But eventually additional students coming into the discussion are only repeating what has already been introduced. A new level is needed. Here is another point of choice at which the instructor can exercise leadership.

The implications of choices at such points as opening the case discussion and use of blackboard should be examined within the dynamics of class discussion. There is no arbitrary standard to establish any one of these choices as better or worse than any other. All of them, like all the multiple objectives, are perfectly legitimate. But, having chosen a certain objective or classroom activity, the professor will then face different consequences.

That class opening which allows for dispersed participation while involving many different points of view leaves the instructor facing a choice somewhere in the middle of the class period. How is he going to get the discussion to converge on one or more subjects in some sort of orderly concerted progression? This critical point in the dynamics of a case discussion might very well be entitled "the point of repetition."

The professor might observe whether the class is adding step by step to a discussion, or whether members of the class are struggling for control to fill the void created by the professor's apparent withdrawal from active leadership. Whether the professor's choice is wise or unwise, good or bad, depends upon his purpose and the way in which he can make use of the consequences. There is a definite role for the discussion leader but many different ways of playing it. Abdicating it altogether would leave a need which other participants would seek to fill according to their relative aggressiveness and purposes — not necessarily for the good of the group.

The professor also has a choice as to whether he will let the sequence of the discussion follow the order emerging in the classroom, or whether he will try to hold it to his predetermined idea of sequence. For instance, in one class a participant asked, "Is there something we have learned about present value of money?" The instructor in this particular class elected to ignore the question of the participant, but 35 minutes later his own questions made it obvious that he wanted someone to bring up a particular subject. No one seemed able to hit upon it. Later in the feedback session it appeared that what he had in mind was "the present value of money," the subject which the student-participant had tried unsuccessfully to introduce 35 minutes earlier. Evidently the class, having seen once that the professor was not interested in the subject, decided not to try that tack again.

This problem of order is often manifest at the point of choice where a professor must shift from the listing of issues or facts to an orderly discussion of the main elements. Experienced case discussion leaders seem to have built up a repertory of blackboard devices and other classroom procedures which help them signal the class that the time has come to make an evaluation of points for further discussion, and to shift from dispersed brain-storming to concentrated collaborative discussion of one point at a time. In the midst of a heated class discussion, this turning of the discussion takes skill and determination. Some members of a class may not be ready. As an indication of the lag between the class and the professor, at this turning point in the dynamics of the class period, were the comments heard more than once during the summer, "You didn't put my suggestion on the blackboard." These comments were usually made by participants who were continuing to play the dispersed brain-storming game when the instructor had, in his own mind, shifted to an attempt at evaluation and elaboration of selected subjects. The instructor and participants were then working at cross purposes. The instructor had not taken into account the momentum of the group.

Detailed illustrations of the points of choice and one instructor's (Professor Paul Lawrence's) selection have been given. Reference to those sequences will show how Professor Lawrence seemed to foresee the need during the class period for some convergence, even though participants had come into the classroom from quite different starting points. Although he allowed men to bring in what seemed most relevant to them, Professor Lawrence seemed to have anticipated one or two significant distinctions or relationships. Thus, without prejudice to the initial free flow, he sorted out the comments, such as characteristics of the new supervisor and of the work group, and noted them on the board in random order as they were volunteered.

When the class got to the point of repetition, he then almost imperceptibly shifted them to thinking about the interrelationships among the categories of comments which had already been made. He was *using* the concepts, and getting the class to use the concepts which he was teaching in the course to add depth to their own discussion. He was effective by timing his efforts at the point of repetition — at the point in the class period when the participants were becoming conscious of the need for a new level of discussion.

Choices Open to the Instructor in Using Tools of Analysis

In the field, the professor-researcher would try out many ways of working with the data. Selecting counterparts to such interaction in the classroom is another point of choice. A principal tool of analysis in class is a comparison of the reactions of participants to a common situation. Differences in interpretation may be explored as access to varied objectives and values.

When a person first looks at a case, there is a natural desire for more facts, but a basic distinction must be drawn between two types of facts. Only the future can reveal some facts to the executive who is faced with a decision or to the researcher who has to write the case from what he knows. It is unrealistic for the student to expect such facts, because these uncertainties are an important part of the decision-making situation. The group discussion provides a rare opportunity to test ideas under uncertainty by observing the varied reactions of other participants. A plan for securing such additional facts might conceivably be a prudent course of action, but the cost of getting these facts, in lost opportunities to act as well as in money, needs to be estimated — a point stressed by Professor Herbert Simon of Carnegie Tech when he met with the Program. For instance, getting an analysis of customers' buying habits, an audit of the treasurer's cash, the relationship of the union leader to the district attorney, might take more time or incur suspicion and more damage than the data are worth. This consideration of what reactions the action of trying to get such data might bring is an important element in the discussion of a case.

There is another type of fact, however, which students rightly expect: the facts they would have if they were in the actual case situation. The executives concerned with the situation would probably know, for instance, how many stories there were in the factory building, how many shifts were working, or what other products were being sold by the sales force. The case writer would have tried to get such background facts if they had appeared to be relevant to the case. If a student feels such facts are missing, their absence need not prevent his using those facts which are available. The instructor might ask the student to make an assumption which appears consistent with the rest of the facts of the case and indicate thereby what difference such facts would make to his decision. Frequently it will be apparent, for instance, that whether there were two stories or three stories in the factory would not basically affect the decision which must immediately be made.

Each instructor has his own analytical scheme which is a major contribution from his course to his students. In the study of a case it is a mistake to think that merely reading a few pages will produce a working understanding of the enterprise. Cases are something like dehydrated vegetables. They need analysis and discussion to bring back to life-like form the situation calling for decision by management. The professor has selected the case because he has found that his analytical scheme is useful. The professor-researcher did "discover" the case in the field and selected the data for the written case; and the case had been approved by the executive as a realistic statement of the situation he faced. But it took a lot of doing. Similarly, use of the case in class takes analysis to bring out the way different dimensions of a situation should be taken into account.

Students frequently make much more progress with a case if the blackboard is used liberally to make clear some of the basic relationships in the case as they come up in the class discussion. The various reactions might then be more easily recognized and compared. Whether or not the instructor asks the class for the material initially or arranges it as it comes out of later discussion, a few structural aspects in the situation usually need to be established. Several uses of the blackboard for this purpose have already been mentioned. It should be noted that some highly successful case leaders rarely use the board because they feel that doing so takes their eyes off the class and they thereby lose clues to emerging insights, and this slows down discussion.

Frequently it helps to get from the class some kind of chart of physical operations or activities. Calling for it when need for it appears in class rewards the student who prepared in advance. The point of repetition often is the place to let students recreate the kind of analytical experience that occurred in the field. The appropriate chart would differ from case to case. In one case it might be the physical layout of a factory, or an office before a change and after. In another case it might be the flow of work through the factory from department to department before and after a change. In still another case, it might be the flow of raw material through various markets and processes on and through various channels of distribution to the ultimate consumer. Getting from the class some such chart of physical operations before and after a change helps to visualize the way different participants saw the activities and how the activities affected interactions. Similarly a quick charting of the *T* accounts might help to analyze the transactions.

Drawing out the pattern of interactions is another way to help the discussion take on focus. An organization chart is one of the most used and misused tools of management analysis. To be realistic, however, the participants need to consider the informal relationships as well as those which were officially recognized. If the instructor, with suggestions from the class, can chart relationships on the blackboard before and after a change, the discussion often will be greatly facilitated.

The activities and interactions of people can be visualized, but their meaning, their value, is more difficult to determine. The instructor may find that it is useful to get on the blackboard the sequence of events and ask what this chronology means to the parties concerned.

By having analyzed the case as a whole in advance, the professor will be able to judge how far and how fruitfully to pursue different possibilities as they appear. He adds to the teaching notes he made at the time of writing the case or selecting it for use each time he prepares to use it again in class. Some of the most effective instructors will spend two or three hours restudying a case before class even though they have used it several times before. The case is different for each group of students.

This chapter began by giving attention to the learning process and the relationship of instructor to student in which learning significant for administration might take place. Selected points have been noted at which a professor's choice becomes part of a total teaching-learning ensemble. Variety in case selection, assignment, class opening, use of audio-visual aids, lectures, closing, and follow-through enrich the process. Although choices at any of these points have consequences which are only partly under the control of the professor,[5] use of cases does not mean that all order is abandoned.

Many visiting professors who had been conscientious in organization of lectures to provide an orderly progression of thought found the apparent disorder of case discussion to be a real block to experimentation with the use of cases. They felt they had so much to cover that they could not take the time to let students grope around trying out their own ideas in ignorance of all that mankind had already developed by way of instruments and institutions of business. Obviously the style each professor developed reflected his own personality and the emphasis he put on curriculum and course objectives. It might be advantageous at this point to reconsider the meaning of order in terms of multiple objectives of education for business administration.

Levels of Class Discussion

Analysis in terms of action involves a different concept of the professor's job from what the visiting professors customarily thought. These differences were highlighted by Professor George

---

[5] *Social Structure and Learning Climate,* by Charles D. Orth, 3rd (Boston: Division of Research, Harvard Business School, 1963).

F. F. Lombard,[6] now Senior Associate Dean for Educational Affairs at the Harvard Business School. Over the years Dean Lombard spoke to the Case Method Program participants a number of times. He had been active in the predecessor of this program and one of those Faculty members most influential in encouraging the inauguration of the program in 1955. The following excerpts, which are particularly relevant to the interpretation of experience in the classroom, are taken from the notes he used in different years.

Inquiry and Decision Making

Two questions that establish the framework for many case discussions are "What's going on here," and "What would you do?"

These questions are a diagnostic question and an action question. They are often held up as the framework around which case discussions proceed.

In my experience few problems develop in the classroom with the first one. That is, if the students will stay with it without getting too complicated, it has a good fit with the classroom. You can speculate about a situation as much as you want. You're removed in a sense from the responsibility of having to take the consequences of a particular action. You don't have to forget this dimension of making decisions in real situations but you can play with it.

I believe there is more of a problem with the second question. Taking action in the classroom in respect to a case is rather different from taking action in the field or in the clinic. Lee Bradford caught the essence of these differences when he said a case discussion is a there-and-then discussion dealing with the decisions that someone else, some other place at some other time makes for real. There-and-then decision making is different from here-and-now decision making. Let's explore these differences for a bit.

The idea I'd like to start with is that talking action and taking action are different.

I am told that the distinction between talking, that is, between words and behaving is a simple one, and that it is a distinction

[6] Although Dean Lombard spoke in terms of progression in class discussion, a similar distinction has also been made by a thoughtful business manager in his analysis of executive operations: *Executive Operations Technique*, by C. Don Williamson (Englewood Cliffs, N.J.: Prentice Hall, Inc., 1963).

that every educated person learns in the course of his college education if not before. I agree with the first but not the second. The second may have been true of everyone who majored in the classics at the turn of the century or before but I doubt if it is true of even most good American college students. Yet the distinction is deeply based in most academic disciplines. One does not think of teaching chemistry by talking about it. One spends hours in the laboratory, one does not think of teaching mathematics by having students read books about it. Students spend hours doing what are called examples. One does not become a doctor by reading books about medicine. One practices medicine under perhaps the most successful educational system ever developed for guiding students from classroom learning to real life situations. One does not learn to ride a bicycle or to float or any other activity of real life, including being married, by reading. One must do. Similarly, for taking action, I believe that it is not enough just to talk about it. At least if one is concerned with efficacy at the point of action one must also do it.

These levels, that of talking about action, which — until I make a correction in what I'm going to say later — is all that one does in the case discussion in the classroom, is not the same as taking action in an organization. Yet it is the latter, the case method training is supposed to be good at. What must be done to make this effective?

You don't escape the problem as I see it by dropping the phrase, taking action, to the more modern phrase, decision making. Talking about decision making and making decisions are different and the difference is exactly the same as between the old levels or rather as between the levels using the old labels which I've already referred to.

The two phrases which I most frequently hear used today in this connection with the case method are "problem solving" and "decision making." Yet I would choose neither of them as being the most accurate. I would select a much less frequently used word, "inquiry." Now problem solving, inquiry, and decision making may easily be conceived as part of a single process.

It seems a logical next step to say briefly how it happened that decision making became the phrase typically used in speaking of the method today. There are several reasons. First, here at the School the method came to us primarily through Dean Donham, who had taken his degree at the Law School. It was the Law School's tradition which we inherited. I am told that prior to and during the early 20s, when these steps were being taken here, the emphasis of the training at the Law School was on the handling of a brief for an appeal. That is, the cases discussed were the decisions of lower courts being appealed to higher courts. In this framework, the investigation of the facts of a case had usually been completed. In this respect, the findings of the lower courts could be taken as given. The important thing for the student was to argue to a decision in the light of precedent on the basis of the known facts and of course in accordance with justice. Thus the method of instruction was oriented away from the inquiry which the lawyer to whom the client first brought the case had to make and towards the decision the appelate judge was being called on to make. At this stage, the procedures of the courts, arguing a brief, winning an argument, what is right and wrong, are important. It is a dramatic setting for students to work in and the dramatic is needed to impel all students to be objective. Yet it is important to distinguish this stage of work in court procedure from the earlier and sometimes less dramatic stages of inquiry in a lawyer's office.

Second, I believe the case itself misleads us in the classroom into thinking that we are primarily concerned with decision making. Let us take, for example, the instructor who not uncommonly has to deal with a student who is reluctant to take a position and argue it. In urging him to do so, with the aim of getting him to use effectively the data that are in front of him in the case, the instructor requires the student to take a stand as if the investigation of the case were complete. The student is asked to deal with what is revealed by the words written on the page. Thus the student may well deem that it is a virtue to decide without making further inquiry even when in the actual situation such additional inquiry might be easily feasible.

The value that this procedure has, I believe, is a product of the teacher-student relationship in the classroom. It is not a value characteristic of the real world. The careful teacher sees that this question is explicated to throw light on the difficult problem of transferring classroom activities to use in practice. But he does not spend too much time at it for the provocative questions of good students can easily trap him — and them — into delaying too long from pursuing with vigor the diagnosis and analysis of the material available.

Thirdly, I believe the phrase "decision making" also persists because it is fairly general in our society that something towards the end of a process stands higher than something near the beginning. Work in a finishing department or at the end of a finishing line is commonly of higher status than the work in the receiving department or at the beginning of the line. Similarly, older people generally have higher status than younger people. Similarly, I think graduate school teachers tend to have higher status than kindergarten teachers, though in terms of who affects the world more, it may well be that the latter have the greater influence and unjustly lesser pay.

In like vein, it follows, I believe, that decision making at the end of the process of problem solving is frequently assumed to be a higher status activity than the more humble and unspectacular phases of investigation near the beginning of the process.

Yet I persist in my view that investigation is the more important activity to problem solving than decision making.

One of the differences between the classroom and the field is that in the classroom the situation is dominated by someone who already *knows.* The act of discovery is what the professor wants the student to experience. This requires fitting action to facts, but is different from the facts. Getting students to make decisions on where to go is different from having them derive a decision on the best way to get where you already know you are going.

From the point of view of a teacher in the classroom, the case method as inquiry is more difficult to explain. As a teacher I frequently find situations in which I feel I know more, and better, about what might be done than my students. I can't dissimulate this. I tell them it is so, and very often what my answers are. Then something interesting frequently happens. What seems so clear and sensible to me frequently does not seem clear or sensible to them. When this happens, it is frequently obvious that a lot that I know or think I know is not understood. The students may not even be interested in it.

What I don't know is why that which I feel so confident about doesn't make sense to my students. With these experiences, the limits and possibilities of the case method as inquiry for me in the role of a teacher open up. If the knowledge I have isn't communicable to others, it isn't truly useful and that to me is a relevant test of good knowledge. Thus, my use of the case method of inquiry in the classroom as a teacher becomes trying to find out what the blockages to communication are that I am encountering. Would another way of presenting or stating the matter be more meaningful than the way I am using? Are there certain conditions which I can help to establish in the classroom which will be more useful?

From the point of view of an administrator I believe that for the future the great need in problem solving is for information.

Problems are more complex; the world is more complex. More importantly than this, most decisions are being made under what are being called conditions of uncertainty. In all these situations it seems to me that the importance of investigation at the early stage of decision making cannot be argued. How else does one deal with what one does not know, with complexities one has not faced before, with uncertainties of the future? By inquiring into them one can reduce the areas of complexity, the areas of ignorance, the areas of uncertainty.

One comes now to a question that investigation without limit leads only to futility. Of course this is so. But in the sense of which I am speaking about inquiry, I have nowhere said, without limit. One must, in business and indeed in most aspects of life, take a stand, make a commitment, reach a decision. One must build a building as General Electric did, a multimillion dollar building for the manufacture of turbines for which they did not know whether there would be a market, if atomic power developed as fast and as economically as it might. One must issue a bond or a stock. These are big decisions of many hundreds of thousands of dollars. Once taken, they are in a sense irrevocable.

There are many other kinds of situations in which it is also necessary to take a stand but without a rigidity that prevents further investigation. I believe the decision, in the sense I'm speaking of it, is one which serves as a platform for further study and investigation of what the future brings. Perhaps there is a time lag before this next stage of investigation get under way. But to think that General Electric will ever stop inquiring as to whether steam turbines or atomic power is the more feasible is ridiculous. They built the building and have continued the investigations and studies.

We need a new statement about decision making in an open society. In an open society, it seems to me, the value of the decision is to close off, at least for the time being, certain things about which no more is likely to be known or in which one, for any one of a number of reasons, decides not to investigate further. Thus, one becomes oriented, not so much to what a company is (a question of being) but what it may become (a question of the future).

Achieving the change in orientation Professor Lombard describes is one of the major objectives visualized in Chapter IV. Bringing about such a change in orientation requires more than just telling the students to change. It comes about slowly through the activities students are encouraged to undertake.

Varieties of Action Questions

Reconciliation of this process of inquiry with decisions on what should be done in a case was frequently perplexing to visiting professors. They were willing to grant that committing themselves to a course of action was an important objective of case study, but questions still remained. "If a professor asks students to get to that point in preparation for class discussion, does he start there or use the class period to go through the whole process of exploration again before asking for commitment?" Professor Lombard was very helpful to the visiting professors in laying out his rationale for dealing with these ever-present questions.

> I have come then to realize that there are five or six different action questions. First, what did X do? Second, what might he have done? Third, what will he do now? Fourth, what would you do? Fifth, what should be done? As I worked over this listing of action questions, I found I could deal with them more effectively by naming them. The first, it seems to me, is *descriptive,* the second deals with *possibilities,* the third is *predictive,* the fourth *personal,* and the fifth *normative.* I'd like to suggest to you that these five different questions then each deals with a different world of thought and discourse and action. Let's take them one at a time.
>
> First, what did he do? For a while, as a teacher, I didn't bother my students with this question. I figured they had all read the case and they all knew what had happened. This was simple stuff, what they should learn in the first year. Yet, rather extraordinary things happened when I started asking them. I found it very worthwhile and much more complex than I'd ever realized. The simple act of observing what a person said to someone else is not easy. In part, I believe we all take what people say, and that they talk, for granted. Being careful to have a clear view of what was said, apart from one's own preoccupations of what one would say in similar circumstances, is never easy.
>
> Of course, at this level, one must be careful that the students don't get away with just reading and repeating the case. I'm asking them to make a selection of the important things and to try to explain why what was said, was said.
>
> Now the second question, what might X have done, opens up the world of the possible. This opens up the alternatives and is, I believe, an extraordinarily rich question which in most cases

needs to be dealt with at length if not exhaustively. Don't confuse this question now with the next one, which is concerned with the future. That is still a different dimension. I'm speaking of what it would have been possible to have done in the situations described in the case. I'm not evaluating too much yet. I'm just asking for descriptions of what it was possible to have done. Let's get outside the framework of the actors and speculate frankly about other possibilities. This is a very big world indeed. It includes merely the practical and the feasible all the way up to the ideal. I like it as a teacher because it is a good testing ground for bringing together the actual and the ideal. You can test these out and build them up step by step. This is an area where you can stick your head in the clouds and still keep your feet on the ground. It's a good posture and good practice for a business administrator. One needs also to have some limits about what is not possible and to avoid speculation unrelated to the real, unrelated to what happened. That also is a useful exercise but not, I believe, in the majority of case discussions. If one can keep the possibility related to the actual, one can take a real step, then another, then another, then another, thus building toward the ideal, whereas ideals unrelated to reality have little relevance in developing the competence of an administrator.

Thus we come to what will he do next? This is the predictive. I suggest treating this question with caution. Most of our cases don't have this kind of material in them. It becomes guessing. We don't have simple patterns of causality in organizations. . . . But a little prediction is worthwhile and certainly good fun and, in some areas where the cases are especially prepared to include data of this kind, it may indeed be well to practice just this kind of question.

Now we come to the personal. What would you have done? This is a much more difficult area to deal with. In the last analysis, a decision should be one that makes sense personally, if in no other way. At the same time, aspects of this decision are not worth discussing in too great length in large classes. The student who takes off easily along these lines is likely to be too sure of himself. He's often out for attention. In other situations, at younger ages than those of the graduate student, he is known as a spoiled brat. He's likely to get into personal assumptions that can't be made available to the class, to rationalize after the fact rather than to arrive at a rational decision. If you get too far into this, you get the showoff. If you reward it too much you get training for having your own way and goodness knows enough executives are oriented this way anyway that they don't need any reinforcement from formal education.

So we come to the last area, what should have been done, the normative. Treat this one with care. There are at least two major directions which it is important for you to have in mind. If one is not careful, one equates the answer to this kind of a question with the personal, the previous one. Yet to make the personal *I should* an imperative is a misevaluation for which most of us are fortunately called to account. Or one makes it a principle, an unvarying rule rather than a guide to action.

Finally, you begin to state the conditions, to clarify the limits of effective action. If this, then that. You begin to cast your language and your thoughts into testable form. Even though you are discussing only one case, you can begin to lay the groundwork for growing and for dealing with many others. I believe this is perhaps the most worthwhile kind of discussion to have in the classroom.

I should point out that all the above questions deal with the level of discussion that I previously called the there-and-then discussion. It is concerned with the action of the people in the case. The here-and-now action concerns not the people in the case but what the students and the professor are doing in the classroom. Sometimes I feel it is very important that time be taken out from the discussion of there-and-then questions to discuss the framework of here-and-now questions. Thus, to discuss the list of questions which we already have in the classroom and to distinguish that from the here-and-now questions is, I believe, an important part of the teacher's responsibility and the fun, at least for me as a teacher, comes from being able to work back and forth from one of these questions to another with my students. As we build these skills over a term, and deal with the relationships of one to another and to real life, both the complexity and the sense of reality of what we are dealing with in the classroom increases.

My answer, then, to the question of what a professor should ask, at least in respect to the action question, is that he should study the conditions under which each of these questions is effective. He should use them all, but be clear to himself and to his students what he is doing and, above all, he should be able to deal with the unexpected — what the framework of questions doesn't catch — when it occurs.

By differentiating six types of questions, Professor Lombard suggested a progression in the discussion of a case that is not necessarily set by the clock. It was based upon the readiness of the class, as the term progressed, to go more deeply into case situations. He approached this gradient in terms of questions from the instructor. He implied that the class was not being left to its own devices.

As noted before, in the Program case discussion leadership rotated within the group. There was some value in the exchange of experiences within the group, but all too often the level of discussion did not deepen. Within the Program there even were one or two case discussions each year which went along very haltingly until the instructor stepped aside at the end of the period for the criteria discussion. Whereupon, instead of considering pedagogy, the class, suddenly freed of "leadership," became so bent upon getting into the case that there never was comment on the instruction — except for a word at the end to call attention to what took place with and without the instructor. In fact, then, questions from the instructor may block learning as well as help it.

"Waffling" was the word the Canadians used to describe what took place in a classroom or among a group of professors when time appeared available which they thought ought to be filled with talking, even though there was nothing more to be said. Nathaniel Cantor would protest that this was part of the "talking, talking, talking" which he held to be the greatest obstacle to communication. The characterization of a discussion depends, however, upon the person making the evaluation. What a silent person thought to be a waste of time, an active, vocal participant might have thought to be a great insight.

Before the Program began, it had seemed that active participation would be an adequate goal. Active participation from visiting professors, who perhaps had never before used cases, surely was achieved. They got into the act. Liveliness of discussion, however, sometimes seemed to become the sole criterion. Dean Lombard indicated by his succession of questions how to go on from discussion to learning. The earlier notes in this chapter, on the beginning and end of case discussions, identified as the point of repetition that critical moving point in the period when the instructor could help turn the varied initial interests and preoccupations of participants to development of the course subject for which he felt responsible.

The categories of questions developed by Professor Lombard

should not be taken as a standard outline for a professor to use in the conduct of every class session. The need for the administrator to discover what direction will help a living enterprise grow, and the need to invent appropriate courses of action, put a premium on spontaneity in the classroom rather than logical order. The professor helps to create conditions for such learning by the selection and sequencing of cases in his course. Some cases lend themselves more readily than others to different purposes. Given the case for the day and the preparation students have actually made, the professor has available to use at appropriate levels of emphasis during the progress of the class period, Dean Lombard's Six Action Questions:

What did X do?	Descriptive
What might he have done?	Possibilities
What will he do now?	Predictive
What would you do?	Personal
What should be done?	Normative
What is going on here and now?	Process

On the Use of Differences

When sequence was discussed, there was no mention of the ending of the class period because it becomes so deeply involved in the use of differences to stimulate continued learning *after* class.

Regardless of how instructors started the discussion, they were surprised at the different things seen in the case by different members of the class. By their second or third turns in leading a case discussion, they had experienced the inherent source of such differences. The existence of such different possibilities in the field was what attracted the researcher in the first place. The support of such possibilities was what the case writer tried to accomplish by keeping out his own evaluations. The case was selected for this point in the course to bring out such differences and to focus attention on the concepts of the course which might be relevant.

Nevertheless, the utilization of such differences in the classroom proved troublesome. Here again in the Program, practice helped; also better selection and sequencing of cases brought up differences in ways which created interest rather than frustration. A "good" case is one which in itself draws a class back to the situation and repays the students who use the ideas of the course by revealing or reinforcing ideas with confirming data.

Tension Between Order and Participation

It is easy to underestimate the strain that case discussion leadership created for a professor who had conscientiously disciplined himself to present a well-ordered exposition of his subject. Likewise, most professors who had not mastered the case method, underestimated the discipline required to set up conditions in a classroom so that *students* could use knowledge of a subject to discover for themselves a living order in the data of a case situation and take responsibility for creating appropriate new directions and means to those ends.

Some professors, in accepting the additional objectives of developing skill and professional identity, and in seeking the associated values of classroom participation, in effect abdicated the role of leadership and left the class members to a struggle for power. Discussion does not automatically insure learning; there must be participation of mind as well as, and sometimes in place of, participation by voice! At the other extreme, a professor might introduce a case; but in order not to risk the loss of his ordered view of his subject, he would then give a lecture on the case. Enlarged administrative capacity derives from the balance of tension between insistence upon the ordered mastery of a subject and the discovery of the emerging opportunities in a living enterprise by a student who is taking responsibility for his own growth. Chapter IV showed how development of a case-course contributes to the student's integrated process of maturing toward these multiple objectives.

The experiment by Professor Lawrence went into detail about what was going on in the classroom. Even though this eight-day period was too short to demonstrate how multiple objectives can be achieved over a nine-month course, there were some clues to the way order can be achieved from discussion of cases. The written case provided the raw material. The situation was selected and assigned because the students in the particular course were thought to be at a point in their development where they could identify themselves with the administrator faced with a need for action. They also were at a point where they were beginning to see the concepts of the subject as tools which might help in dealing with the administrative situation.

Many visiting professors found it very difficult at first to believe how much work was necessary to move from the raw material in the case to discovery of the enterprise as a living organism struggling for survival in its environment. However, until the situation takes on reality, attempts to arrive at a decision or a course of action for it are abstract and unrealistic. The student has to make the same kind of analysis the professor-researcher made in the first place. He needs to visualize the work being done, who is doing it, what it means to them, and the transactions between the company and its suppliers, workers, and customers. He needs to think about how all these factors were being changed. This is what is meant by recreating in the classroom the experience the professor had in the field.

In a classic case, the narrative stops with an unresolved issue requiring decision or action. Just as the researcher had to separate out his own reactions to the facts in order to weigh the meaning of alternative actions to individuals in the situation and to the organization as a whole, just so the student has to differentiate between his own reactions to the situation and those of actual participants in the situation. All this is far more easily said than done, because analysis of a case is a quite different activity from the reading of most other documents.

As Professor McNair said in his talk about the written case, there is a big gap between the practicing administrator and the student who has to comprehend the situation. This is particularly true of students at the undergraduate level. The selection and sequencing of cases progressively help to close this gap. The opening case at the Harvard Business School for many years was concerned with a particular firm which made student diaries. It was relatively easy for students to comprehend the obsolescence of diaries after the calendar year had started. They also could see that some estimate of sales had to be made well in advance of the opening of the school year in order to produce diaries efficiently in time for the selling season. In addition, they could see that this production and inventory policy took money.

On the job, the executive cannot separate considerations into arbitrary subjects for abstract and isolated consideration. He must deal with the total situation as he finds it. This was one of the reasons why actual cases were introduced. But the instructor must recognize that if he takes a cue from Nathaniel Cantor and asks at the start of the class, "What interests you in the case?", a participant may not feel interested in the particular subject in the course for which the case has been selected. There are many elements in most situations even though at least one of the critical elements is related to the subject currently before the class. The instructor could show where he wants the class to focus by starting with a question about pricing, for instance. Wherever the discussion starts, the case provides an opportunity to relate the thinking of the participants about the subject in general to at least one specific situation in which the subject appears as one part of a complex living whole. Therefore, sometime before the class is over, the personal relations of merchandise manager and operations manager, for instance, could be explored as an aspect of the particular pricing issue, even though it might not be at exactly the time anticipated by the professor.

Cases were generally selected to provide participants with situations in the classroom in which they could foresee the need for some action. In the Middle Management Program, the tendency at first was for the men to speak immediately with confidence about the action which occurred to them, based upon their past experience, admitting little scope for discussion. The cases, however, had been selected because the manager in the actual situation had felt such conflicting pressures that he was concerned about which action would be most effective, and other participants would present one of these alternatives with just as much assurance based upon his past experience.

On-the-job prompt action is imperative, but one of the reasons for the formal study of administration is the opportunity for students to reexamine their own habits of action. The task of the participants in the discussion, therefore, particularly the instructor, was to suspend their initial impulses and to take advantage of the occasion by exploring alternative ways of sizing up or analyzing the situation. For a class of management participants, the instructor may need to devise ways to get attention to analysis, while with an academic group the shift from analysis to action may need special attention.

Participants and particularly the instructor gradually became aware that their own attitudes frequently were being projected into discussion of case events. Often at first reading, participants had little sympathy with the man who provided the case, but the experience of heated disagreement within the group made the perplexity of the executive on the job more understandable. One purpose of the case was to give students an opportunity to test and modify their thinking and reactions about a subject in relation to a specific situation. Therefore, a major factor in the case discussion was the personal reaction of each participant. By recognizing impulsive tendencies or obsessive preoccupations, each participant gained personal insight as well as greater capacity to ask discerning questions. How the instructor recognized and chose to deal with these aspects of case discussion was considered frequently throughout the Program. It was part of "learning which makes a difference."

Decision making by management is a unique mixture of art and logic. Effective managers are identified by their ability to ask discerning questions as well as by the stereotype of action in a crisis. These questions seem to be prompted by an understanding of the enterprise as a living whole and what it needs for growth in a constantly changing environment. The varied reactions exposed in group discussion, give the instructor the opportunity to demonstrate the dynamic forces in adaptation of an enterprise and its environment.

Chapter III stressed that the written case should be limited to a sampling of the data. In the classroom these data are what stimulate the students' involvement. The need for acting arouses their identification with the role of the administrator in this situation, as well as their discovery of effective purpose and direction from skillful thinking about the situation as a whole. Use of the case also shows what knowledge students bring to this case, and take from it: a more effective generalization about the subject of this course, more logical inference, and increased maturity. Here in the classroom the task is to re-create with the written case the experience of discovering the action point which the researcher had discovered in the field.

As indicated above, such analysis would reveal that there is more than one interpretation of the situation which a reasonable man might make. Management is a creative act in which there is room for innovation. The facts of the total situation must be taken into account. Management must exercise its creative powers within the limits of a set of forces which often are conflicting.

This discipline of the specific situation tested the generalizations arrived at by group discussion of a subject. It is this discipline of dealing with a specific situation which later helps students to put their ideas into practice on the job. Thus the tension between subject and situation is the motive power which the professor utilizes in leading a case discussion. He uses his authority to help students distinguish among theories and facts in the case and their own personal values.

Leap and Logic in Learning

Practically all of the discussion thus far has assumed that the instructor would be in the role of questioner, if he were not in the role of lecturer. Dr. J. Richard Suchman introduced participants to a different role, one in which experimentation might well prove fruitful. In a sense he would use the fact that, after careful preparation of a case, students almost inevitably would come to class with courses of action which were distinctly different. If he were to teach business administration he would reward the students' preparation by allowing each student to state briefly the course of action he would choose, until the discrepancy dramatically demonstrated to the class the need to reexamine the data in the situation. Dr. Suchman held that the plan of action was in the nature of a theory, to which one inevitably makes an intuitive "leap." The task of the instructor then is to invite attention back to the data in order to test the adequacy of the theory.

Such an approach would reward the student who had spent time in careful preparation and would demonstrate the utility of mastering the conceptual tools of analysis and, at the same time, provide for an exposure of the conditions of the case early in the class discussion.

Somewhat surprisingly, Dr. Suchman, one of the most helpful and stimulating speakers in the Program, came to attention because of his experiments with junior high school students in the teaching of "inquiry." At the time of his talk, he was on leave from the University of Illinois, and was serving as Director, Curriculum and Demonstration Program, Office of Education, Department of Health, Education, and Welfare, in Washington, D.C. He made presentations to the Program in 1964 and 1965, outlining the results of his experiments in promoting what he called "autonomous discovery." A particular contribution of these experiments to classroom conduct was Dr. Suchman's explanation of devices for making the thinking of a student articulate and thus available to the instructor, to himself, and to other students. These devices were similar to what Dr. Melvin T. Copeland called "getting the student to reveal his blind spots" by case discussion.

Dr. Suchman conducted his inquiry training by confronting the junior high school students with "a discrepant event." He used short motion pictures for demonstrations, primarily in natural science classes. A fluid would go uphill, for instance, when the technician put her hand around the bottle in which it was stored. Dr. Suchman would then accept questions which could be answered by "yes" or "no." In his terms, he was providing the student with a responsive environment. By limiting the instructor's participation, he drew out the students' theories in a form which could be considered by the instructor and by others in the classroom. This arrangement also allowed the student to test his theory by suggesting experiments and getting a feedback. Dr. Suchman was getting the student not only to discover a theory for himself but also to test its utility. He helped participants to clarify the "leap" from data to theory and the "logic" needed to test theory with data.

In an analogy using Dr. Suchman's approach, the case facts present to the student a "discrepant event" at the action point. The equilibrium between the enterprise and its environment was somehow upset or about to be upset. It is alertness to such points of action and the discovery of useful interpretations of the events that lead to appropriate action. In brief, the use of a case helps to place responsibility for inquiry on the students. It is here that the difference between the instructor's questions about a subject for discussion and the issues in a case requiring action may be seen.

Conduct of the class by the case method gives the professor access to the students' thinking and use of the subject, and thus allows him to gauge what learning has taken place. This is a more functional guide to progress in the course than mere coverage of the subject matter. For this reason, clarification of objectives and careful sequencing of cases and concepts are essential in developing the course. Thus the instructor can be paced within the class period by the learning that emerges in that class, knowing that this will enrich and be fulfilled by subsequent case discussions.

Such progressive utilization of insights from one class period to the next was difficult to accomplish in the Program because the instructors were rotating. It also was difficult because visiting professors had in mind the widely different attitudes toward such learning which prevailed in their own institutions.

Can a University Foster the Creative Leap?

Both the scientist and the artist need "leap" and "logic" — the alternation of vision and analysis — as does the administrator. The scientist may at least remain respectable if he is skilled in analysis, whereas neither the artist nor the administrator in a rapidly changing society can exist without some measure of success at creation.

Whether or not the essential art of management is part of the curriculum, or of a particular course, is more a function of the university's capacity to provide a congenial atmosphere for artists in its faculty than a deliberate evaluation of management as an art. One of the most sobering perplexities of the Program was the repeated questioning of the extent to which the conditions of creativity could be maintained within a university. A statement of these conditions by Professor Carl R. Rogers frequently was found to be a useful focus on the issue.

> What are the conditions within the individual which are most closely associated with a potentially constructive creative act? I see these as possibilities.
>
> (a) *Openness to experience: extensionality.* . . . It means lack of rigidity and permeability of boundaries in concepts, beliefs, perceptions, and hypotheses. It means a tolerance for ambiguity where ambiguity exists. It means the ability to receive much conflicting information without forcing closure upon the situation. . . .
>
> (b) *An internal locus of evaluation.* . . . The value of his product is, for the creative person, established not by the praise or criticism of others, but by himself. . . .
>
> (c) *The ability to toy with elements and concepts.* . . . From this spontaneous toying and exploration . . . arises the hunch, the creative seeing of life in a new and significant way. It is as though out of the wasteful spawning of thousands of possibilities there emerges one or two evolutionary forms with the qualities which give them a more permanent value.[7]

By the end of eight weeks, participants began to see that discovering a concept for themselves and playing with it in a number of cases might have something to do with administration. In Dr. Suchman's terms, the discovery of corporate purpose was a "leap," not a process of "logic," although it had to be rigorously tested. If the determination of policy direction were taken as a major function of the administrator, then somehow he must find ways to give men courage to leap. But this seems very difficult to do in an academic community. There is little to show for it — a residual increment of courage, skill, judgment — nothing to publish. There is a similarity between the search for

---

[7] "Toward a Theory of Creativity," by Carl R. Rogers, *Creativity and Its Cultivation,* Harold H. Anderson, ed. (New York: Harper & Row, 1959), pp. 75–76. [Author's italics.]

policy direction and the fundamental process in evolution stated by the French scientist, Teilhard de Chardin.

> . . . We find the fundamental technique of *groping*, the specific and invincible weapon of all expanding multitudes. This groping strangely combines the blind fantasy of large numbers with the precise orientation of a specific target. It would be a mistake to see it as mere chance. Groping is *directed chance*. It means pervading everything so as to try everything, and trying everything so as to find everything.[8]

The result of autonomous discovery, of directed chance, is called purpose. Given purpose, the rule of reason then tests all means by rigorous logic against this end. These later activities are respectable in academic circles, but where in the curriculum is there room for autonomous discovery, for groping — directed chance; for experience with the evolution of direction?

The acknowledgment of art as essential in administration and its neglect in the curriculum can be illuminated by the comments of a resident visiting artist in the University, who wrote:

> Perhaps we may move again toward that antique and outmoded ideal — the whole man. . . . [his] . . . need to make desperate choices among his own values and his wants, the need to reject many seeming benefits or wishes. For it is only through such conflicts that his values become sharpened. Perhaps it is only through such conflicts that he comes to know himself at all.
>
> It is only within the context of real life that an artist (or anyone) is forced to make such choices. And it is only against a background of hard reality that choices count, that they affect a life, and carry with them that degree of belief and dedication and, I think I can say, spiritual energy, that is a primary force in art. I do not know whether that degree of intensity can exist within the university. . . .[9]

If this essential creative force is to be studied and communi-

---

[8] *The Phenomenon of Man,* by Teilhard de Chardin (London: Wm. Willins & Co., Ltd., 1959; New York: Harper & Row, 1959 — Harper Torchbook Edition, 1961), p. 110. [Author's italics.]

[9] *The Shape of Content,* by Ben Shahn, New York: Vintage Books, Random House, 1957 (reprinted from Harvard University Press, © President and Fellows of Harvard College, 1957), pp, 9, 13.

cated in a university's school of business administration, it is imperative that the faculty be clear about its responsibility for the professional objectives and attributes of the student as a whole person. The study of administration, like the study of art, is not an easy thing to incorporate into the ordinary academic community. The case method pattern includes a number of inventions from past trials and error which, taken together, provide a way of bringing this essential element of administrative orientation, skill, and policy direction into a community where important analytical skills and generalizations about human institutions also thrive.

Chapter VI

Alternative Methods or Alternative Objectives?

CURRICULUM AND CAREER QUESTIONS IN THE CASE METHOD COMPLEX

DISCOVERING A CASE SITUATION in the field, preparing the written case, integrating it with other cases and concepts in a course sequence, and involving students in class discussion almost completes exploration of the case complex. The final pervasive element in the set of variables is the particular academic community which brings students and professors together. University administrators faced with the problem of financing facilities for an exploding student population must dream of the idyllic log bench of Mark Hopkins. The interdependent relationship between the curriculum which attracts the students to the institution and the career prospects which attract the faculty members is a dominant influence shaping the case method pattern, or any other pattern, in different institutions. The discussion of a case in class is not only in the context of a professor's course. It also is in the larger context of a curriculum and the particular institution. Visiting professors would on occasion comment that what they were experiencing during the summer was "unreal."

But the discrepancy between the case method pattern in the summer program and the situation in which they were imagining themselves back home was not entirely a matter of "cases."

Throughout the Program the cross-section of participants from various schools would compare how the conditions of the Program differed from those at home. As each summer session concluded, the visiting professors became more and more aware that conditions in their own universities would not enable them to concentrate as fully on students and case-course development as they had in the Program. They realized that in each of their institutions there were different pressures which would modify the mixtures of activities they were experimenting with in the Program, and they also questioned how career and curriculum issues were dealt with at Harvard. Almost from the beginning, therefore, and throughout all 11 summers, visiting speakers were invited from differing academic institutions to consider with the participants how these differences might affect the case method pattern.

In the operations of the program discussed thus far, the professor could clearly see himself as the actor. He could try different approaches and get the reactions of the group. But when it came to curriculum and career questions, the individual professor felt acted upon. He felt constrained by departmental organization and standards for selection and grading of students, as well as by promotion criteria for faculty.

The visiting professors in each group always included an amazing cross-section of people representing a variety of personal experiences in a wide range of universities in the United States and elsewhere. Informal contrast and evaluation went on continuously and broke out into open discussion during almost every formal meeting in one way or another. In these discussions it became clear that the individual professor's interest and proficiency developed in relation to the institution which had brought him and a group of students and other professors together. What attracted students to the school and what determined a professor's promotion could not help but influence course objectives and methods.

The Program was organized on two basic assumptions: that

professors were interested in the study of business administration and that their students were interested in becoming administrators. For the activities of the Program, these assumptions largely prevailed and allowed creative involvement in all phases of the case method complex. However, these assumptions did not prevail everywhere.

Many questions originally directed at the case method turned out to derive from basic differences of opinion about business administration and the study of it in a university. They were questions about the value of studying business administration by any method. Outside speakers, with a wide range of viewpoints, were invited to present their views to the group. Taking issue with the case method publicly made criticism legitimate and helped participants clarify their personal and institutional positions.

After the participants returned to their home institutions, their follow-up letters, phone calls, and visits revealed in increasing variety the institutional determinants of the case method pattern. This information was used in selecting speakers in successive years. Fortunately, The Ford Foundation encouraged the administrators of the Program to keep in touch with the alumni and to consider more fully the "transferability" of the case method. Clearly, the case method in a special program such as this Visiting Professors Program necessarily differed from what one would find under different conditions and in other universities. It is important to be aware of these differences.

The value of exploring these differences while participants were together was part of the original design of the Program. And, to accomplish this design, in the first Program, Professor James W. Culliton was secured as a coach. He was then director of a special program for the study of business administration by undergraduates in the University of Notre Dame. In the 1962 Program, Dr. Ike Harrison, Dean of the M. J. Neeley School of Business at Texas Christian University, served as a coach. He was experienced in an undergraduate program in business administration and had been a participant in the 1958 Program. Other members of the coaching staff throughout the several Programs had also had experience in teaching in various schools.

In addition, the formation of the Intercollegiate Case Clearing House, shortly after the beginning of the Program, provided active discussion with a cross-section of deans from other schools of business administration. The Advisory Board of the new Intercollegiate Case Clearing House was made up of the Executive Committee of the American Association of Collegiate Schools of Business. Typically there were 12 members of this committee, including the officers, and approximately 3 new men came into the group by rotation each year. Many of these deans had nominated participants in the Program and had themselves participated as speakers. Thus, over the years, the Program had the benefit of reactions from more than 40 deans of other schools of business administration on some of the major concerns of continued case development and use. Also, from year to year, the program direction benefitted from the increasingly detailed view of demands which different academic communities made upon participants and of the support available for their interests.

Thus, each summer, participants were able to examine the wide range of attitudes toward the study of business administration prevailing in the institutions represented, plus those in a variety of other schools from which speakers were drawn. (Appendix 3 lists the guest speakers.)

Some of these speakers have already been referred to in connection with field work, case writing, course design, or case discussion in the classroom. Unfortunately space does not permit inclusion of all they did to demonstrate the tremendous complexity of the task of education for administration in business; nor is there space to list the exciting possibilities which could materialize when individual professors or institutions mobilized their resources to accomplish some part of the task. Since many of these outside speakers came from schools considered to be skeptical of the case method, they made it easier for participants to express openly their questions about limitations of the case method and to consider alternate approaches. One will note also that guest speakers also included well-known textbook authors; theorists of organization; practicing executives; philosophers; experts in learning theory, achievement motivation, student selection, and social structure; deans of undergraduate schools; deans of graduate programs; and presidents of the American Association of Collegiate Schools of Business. Some of these speakers were concerned enough about the whole task of business education that, even though dubious about the case method, they not only came to deliver their talks but to visit work sessions even for as long as two or three days. The resulting opportunities for informal discussion helped to sharpen the issues and open the possibilities of experiment. Especially helpful were the deans and former Program participants who came back and discussed frankly the criticism as well as the satisfactions which came from their experiments in case-course development and curriculum revision after they returned home. The more accurately participants could see the similarities and differences between the conditions surrounding the Program and the conditions under which they would work at home, the greater was the prospect of creative adaptation of a personally workable case method pattern.

Each speaker, regardless of his orientation, would insist somewhere in the discussion that his school was open above all to academic freedom for the individual professor to study and teach as he saw the truth. But when he described the program as a whole and answered questions about student selection and faculty promotion, limits to freedom could be seen. Whether a speaker's comments line up with academic freedom or a school's party line often depends upon very small differences in point of view. Each outside speaker left the group with a parallax effect. When a person looked at him with the right eye, having the left eye closed, he could see the speaker clearly free to follow his personal interests. But when a participant looked with the left eye, having the right eye closed, he could see the speaker in line and conforming with a carefully integrated program. The inevitable contrast gave depth and perspective to the stresses inherent in an academic community.

Chapter IV revealed the need to examine course objectives from several different angles. It was seen that the design of a course is a function of the comparative weight given by the professor to orientation, skill, information, concepts, and growth of the individual. The attributes to be developed are multiple and compound:

Knowledge and the capacity to generalize
Both policy direction for a specific enterprise and the logical system of the particular academic subject.
Skill
Both for discovery of goals from complex situations and for logical inference under given conditions.
Orientation
Both as a responsible leader in action situations and as a detached searcher for understanding.
Maturity
Both in control of impulsive action and in avoidance of obsessive overthinking.

It became clear in the talks by visiting speakers that the relative importance of these values differed from school to school as well as from professor to professor. The weight given to these different values in the school as a whole significantly influenced how far an individual professor felt he could stress administrative orientation or any other value in his course. Some schools are organized to be more congenial communities for those who want to study administration in business than are others, just as some universities are better places to study chemistry than others.

Few, if any, would deny the need for the above attributes. More would question the emphasis which should be given, or which in honesty they could give, to some of these objectives. Many would differ on the extent to which the faculty should try to develop all these attributes rather than let the student take the initiative in drawing what help he could get from various professors and ultimately from his associates in business. Over the eight-week Program each participant could explore these differences not just as points for argument, but as ideas with which to experiment in his own course and with his colleagues in making the curriculum more effective for students.

The Program attempted to show how faculty and students could develop all these multiple and compound attributes as interdependent qualities of a living person. After eight weeks the participants protested that the term case method as they had initially known it was too narrow. Some outside speakers similarly seemed to have a mechanistic view of the case method as a device, superimposed on what a professor would have done anyway. However, the wide range of viewpoints brought into the Program enlarged the participants' awareness of the ramifications of education for administration in business, even though none of the subjects introduced could be dealt with exhaustively.

This chapter will first suggest some of the distinctions highlighted by the succession of outside speakers. Later, the chapter will report the experiences of Program alumni as they sought to follow through under the varying conditions in their own academic communities. Further experience is discussed in subsequent chapters and appendices.

Exposing Participants to Different Emphases

In the discussions, various emphases recurred concerning:

Basic academic disciplines
Research
Professional development
The art and science of administration
Practical experience
The social structure of academic communities

As the pluses and minuses presented by speakers from different schools were discussed, straw men vanished. In their place came awareness that each participant and his colleagues at home could design their own patterns of education for business administration. The persistence of differences over the eight weeks of the Program helped everyone to realize that multiple objectives were much easier to state than to achieve, and increased the participants' appreciation of those points which some other professor or school might choose to emphasize.

On Basic Academic Disciplines

One of the stereotypes most frequently contrasted to the stereotypes of the case method was the emphasis on basic academic disciplines. On several occasions Dr. Allan Wallis, then Dean of the School of Business at the University of Chicago, met with participants in the Program and stressed, "It is the job of the faculty to teach theory which is not dependent upon circumstances but systematized and enduring." In line with such a point of view, the curriculum is organized first around basic science disciplines — statistics, econometrics, accounting, how judges make up their minds, microeconomics, macroeconomics, and behavioral science. Such a program would then be followed by exposure to functional fields and business policy. Corollaries of such curriculum planning are the beliefs that "the best research will be done by social science based faculty members," and that "the case method is only the culmination of the rudimentary stage of professional education for business administration — the apprenticeship method." "There is a place in the Business Policy course at the end of the program for cases; because we don't know much about situational thinking in the social sciences." Such points of view were obvious grist for the mill of discussion in the Program where men were devoting 24 hours a day to "situational thinking" as a cornerstone of the discipline of business administration. What was at issue in these fascinating discussions was not the case method but the weight to be given to the multiple and compound objectives outlined at the start of this chapter.

During the period that The Ford Foundation was supporting the Program, it also was encouraging schools of business to draw more fully upon the behavioral sciences. Chapter VIII tells of the experience which alumni of the Program from Southern schools had in developing business policy cases in collaboration with behavioral science consultants. While the eventual benefits of cooperation were great, the experience showed how very difficult it is for men under one discipline to work in another.

On Research

A second and related emphasis was upon research. Because the faculty of business administration cannot avoid responsibility for the multivalued ends of men and institutions, the school of business administration cannot be content alone with what science has demonstrated. Administration must deal with the determination of purpose under unique conditions. Bringing both dimensions, purpose and science, into the curriculum complicates the challenge so often leveled at business schools in the last decade. Few would argue against the proposition that business education will attain its rightful place in the educational world only as it pioneers in research.

Many of the outside speakers concerned with research called attention to the rapidly developing interest in more theoretical literature, especially development of a theory of administration and decision making. One speaker asked bluntly, "Are case writing and a theoretical orientation incompatible?" He added the comment, "Unless it is feasible to generalize from experience, then the study of unique situations has no meaning." Participants in the Program that same summer (1959) had seen for themselves the complexity of generalizing about a policy direction for a particular company from the facts in the concrete situation, even when research generalizations about such situations in the abstract had been well documented. For instance, the first two days of the experiment in merging cases, concepts, and research findings of "Management in a Social System," which were reported in Chapter IV, illustrated the sobering ratio of cases to concepts needed in order to orient professors to the use as well as to the intellectual comprehension of a generalization. Most of the participants had known the Hawthorne findings, but did not easily recognize opportunity to use the concepts in case analysis and action.

For the years 1964 and 1965 the Research Workshop in Finance met concurrently with the Program. The coexistence of these two groups further illustrated the difference between the needs of professors seeking uniformities among companies and the needs of professors (even the same professors) when seeking to understand the goal setting, policy-direction-reaching efforts of a particular company. Both are important, but they differ. Increasingly, business executives as well as professors recognize these differences, and are willing to try to understand the language required in order to be effective in each orientation. Incorporation of research findings into a course and into a curriculum brings attention again to multiple objectives and to the responsibility of a faculty for development of all these professional attributes in students.

Although the Finance Workshop participants and the Program participants came from similar schools of business administration, lived in the same dormitory, and ate in the same dining hall, there was strikingly little professional interaction between the two groups. As Director of Research at the Harvard Business School, Professor Bertrand Fox, the Director of the Finance Workshop, spoke to these as to previous Program groups. The participants were much interested in his outline of research policies and projects. Advance planning for both groups included arrangements for a few joint sessions. But at attempted interchanges between the two groups, there were polite pauses and an intellectual acknowledgment of overlapping interests. The participants might say, "Let me have the results of your survey," or "I'd like to use your case." Although socially they were friendly enough, each group had its own preoccupation. The difficulty of shifting from the project research reference to the case-course reference and vice versa cannot be dismissed lightly; both these groups were uniquely free of other distractions.

If a business school graduate who is now a chief executive were to come back to his school for advice, even the most up-to-date knowledge of the faculty members would not relieve him of many of his present perplexities in trying to integrate all these developments into the particular enterprise for which he has responsibility. Goals and directions for an enterprise are quite different from the knowledge needed in the various systems of an enterprise which somehow must be related by the emergence of a harmonious set of business policies. In his enterprise, the business school alumnus is faced concurrently with a number of cases, each of which would challenge the research efforts and course concepts of a different professor. Each professor in turn would be seeking the comparative experience of many executives in order to test his academic research models. And yet the researcher would ultimately turn to the executive and ask: "What's your purpose?" "Where are you trying to go?"

This troublesome junction between the purpose of the particular enterprise and the research model in the study of business administration ran as a "leitmotif" throughout all 11 years of the Program.

On Professional Development

Participants, therefore, always were stimulated by contrasts with administration and education for law and medicine, where action and expanding knowledge also had to be merged. Thus professional development appeared as a third emphasis. As Professor Russell D. Niles of New York University Law School said, "In a profession the only fatal error is paralysis of action." "It is a job of the law school to accelerate men toward professional maturity." Students buy casebooks. Each student then, in effect, writes his own textbook from the study and analysis of the cases. But the profession needs more. Professor Kingman Brewster, who later left the Harvard Law School and subsequently became President of Yale University, also described the period from the 1880s to the 1930s as dominated by Noah Langdell's belief that there is "a discoverable logic in the rationalizing and distinguishing of a grouping of cases." But in the 1930s the lawyer emerged in the role of counselor. He became a lawmaker in government agencies by exercising administrative discretion. This new role did not lend itself to the traditional case method of the law school. Therefore, Professor Brewster saw the law schools turning to "fact problems" with new pedagogical questions, which echoed the questions so often raised in Program discussions, "How make these fact problems manageable and still realistic?" "How bring into the Law School classroom counseling, negotiation, building relationships, and use of experts in anti-trust and international relations?" "How maintain rational content in open-ended discussion?"

Education in the medical profession provided another view on the increasing stress between the handling of cases and the pursuit of research. In the clinical conferences, experienced practitioners who came in to diagnose the case were becoming embarrassed by the critical analyses of students, who based their diagnoses on laboratory test results and theoretical models. Dr. C. S. Burwell, Assistant Dean of the Harvard Medical School, and Samuel A. Levine, Professor of Medicine, obligingly met with the Program and subsequently introduced Dr. Bert Vallee, Director of the Biophysics Research Laboratory. They both encouraged participants to study the relationship between older and newer methods of medical education and how such evolution might develop in the study of business administration. Dr. Bert Vallee challenged the case method of teaching in a professional school. He questioned the process of classifying clinical syndromes, and favored putting energy into the development of biophysical models and the analysis of chemical, physical processes. He made everyone feel that they understood the biochemical model of gout which he used as an illustration. Also all recognized the tremendous gains in health which have come through the laboratory discoveries of penicillin, antibiotics, vaccine, without which the medical practitioner today might not be any further advanced than the practitioners of 50 years ago.

Even the medical school professors, however, were far from clear and far from agreement about what these developments meant for the curriculum. Nevertheless, possible analogies with business administration and related sciences always stimulated vigorous and continuing discussion in the Program. Dr. Vallee denied any knowledge of business administration and, in the discussion contrasting medical and business research, pointed out that each cell in the body includes the complete chemical program of the entire body. It was on this fundamental organic characteristic that he and his associates were developing models of the various mechanisms within the body for dealing with these chemical programs. Participants in the Program recognized that in the business firm there is a significant difference because an explicit program for the entire enterprise rarely exists anywhere, and there is tremendous distortion in communicating whatever pattern there may be from unit to unit and from level to level in the organization. A further difference appeared in the problem of goals. Dr. Vallee acknowledged that even the most ultimate knowledge of biophysics will not tell the patient where to go and what to do with himself in time, to say nothing of eternity. The interrelationship in the curriculum of medical schools between the teaching of medical practice and research in the development of medical sciences has some significant differences as well as thought-provoking parallels for schools of business administration.

On the Art and Science of Administration

Again, the question is not whether to stress the development of science or the art of policy direction. Both are needed in administration. But it is difficult to discuss multiple levels. Thus a fourth emphasis was on what might be called the art and science of administration. Dr. Leon Bach, who in 1955 was then Dean at the Graduate School of Industrial Management, Carnegie Institute of Technology, spoke to the first Program using the title "Limits of the Case Method." He had been invited to speak to the group because in the preliminary negotiations about the Program with The Ford Foundation he had felt so strongly about the need for coupling the study of cases with the construction of models. In correspondence about his first talk, the development of the Program, and subsequent visits, Dr. Bach wrote:

> We are disagreed to a considerable extent, I suspect, on (a) what is the most effective way to teach this art, and (b) the optimal role of research and its relation to teaching activities.
>
> On (a): To say that administration is an art is not at all, in my mind, to imply that there is not a large amount of solid and essential theory that underlies it. I am sure that engineering, for example, is really an art, but it rests very basically on the physical sciences and mathematics. All the evidence is that the engineers with the most solid foundation of this sort are by far the best "artists" in their field. In painting, to take a quite different kind of example, I am told by both the teachers and a few very distinguished painters whom I happen to know that the story is much the same. The first-rate painter must be perhaps in some sense a born artist, but even the genius must assimilate a solid and substantial, disciplined body of knowledge on techniques in painting. He may be able to develop this for himself, but almost no one can and even that rare individual is apt to lose a vast amount

of time that he could have saved if he has to relearn everything for himself without any reliance on the accumulation of past experience of others. Just so in administration, I think. My attempt to break down the objectives of training for administration into the sub-objectives that I spelled out rests on an attempt to break down into some of its central parts the art of administration, just as we can break down the art of engineering or the art of painting or the art of anything else, I believe, into some subcomponents. This attempt does not imply that the parts are separate or that they are more important than their aggregation into the whole.

I am aware of the danger that by specifying the sub-objectives and that by verbalizing the concepts and theory implied, we may somehow dull the "insight" and "imagination" that are essential for the really first-rate administrator. I am quite prepared to believe that theory and analytical concepts, wrongly taught, can have a very stultifying and narrowing effect on a man's thinking. You may recall that I was emphasizing rather heavily at Harvard something I called "search" behavior in managerial thinking; this is precisely an attempt to get away from having people feel that all they need to do is to slap a formula on any problem. But I argue that basic theory and analytical concepts, properly taught, can be very helpful in creating this over-all behavior that we loosely call the art of administration.

On (b): We are talking primarily about best paths to an improved art of administration for twenty-five or fifty years hence. Here there is plenty of room for people to suggest other research approaches. I am pretty thoroughly convinced that the path we are choosing is substantially the best one in the long run, and I think I can produce some rather impressive evidence by analogy with other fields, but this is not to say that I suspect that other people will, or even ought to, agree with me. If my general argument is right, however, the advance toward more effective training for administration is going to be along the lines of whittling down the unknown, the intuition, and the "hunch" and increasing the quantifiable, the carefully measured, and the "known." This is going to take a long time. I think it is a movement in the right direction and that it ought to be encouraged and pushed. But neither you nor I, I suspect, will even come close to seeing the day when managerial decisions can all be made on any such basis.

Fortunately Dean Bach was able to stress these points in meetings with the Program in other years. His colleagues and men from other schools with kindred interests in model building, simulation, and games gave participants an opportunity to consider the state of development of these and other means to give students acquaintance with the nature of an enterprise.

On Practical Experience

Persuasive experts pointed to the growing technology and the responsibility of universities to give students an adequate background. The case method was thus criticized as failing to equip a student with the rounded background of currently significant information needed to start work in business. Although some critics of schools of business have pointed out excesses in proliferation of courses on specific industries, the values that come from such faculty organization and such course selection by students must not be overlooked. Surely the consulting opportunities which come to faculty members who have specialized knowledge of a particular industry are persuasive influences in the choices that individual faculty members may make in their own research plans. Business administration has always been dependent upon relevant information. Commercial intelligence has always carried a high premium in the conduct of practical affairs, and there is a market for the advice of experts. But there is more to communicating the facts of administration within a university than giving the expert an endowed professorship.

Each year the Program included a few professors who had had active careers as business administrators and who wanted to obtain experience from the Program as part of their transition into the academic world. They were particularly dissatisfied with the gap between the academic disciplines, abstract subjects, and the realities of the administrative experiences through which they had lived. Active business administrators sharpened these continuing discussions. For instance, Dr. Edwin Land, then President of the Polaroid Corporation, thought out loud with the participants about the way in which modern science could be utilized to satisfy the deep needs of man. He was sure that enough was now known about the processes of learning to justify undertaking commercial commitments which involved filling gaps even in basic research knowledge. Management was thereby challenged to organize in such a way that production workers as well as scientists could be drawn into the creative process. It was not enough to know about things in general. There must be a way of knowing "precisely." Dr. Land's visit underscored the observation in Crawford H. Greenewalt's book, ". . . the basic requirement of executive capacity is the ability to create a harmonious whole out of what the academic world calls dissimilar disciplines."[1]

Participants were impressed with the range of exposure and understanding of academic disciplines of industrial executives whom they met in case research and as speakers. The educational program in which business organizations were involved from the top to the bottom, demonstrated the reality that with accelerating change, learning is a never-ending process. As one executive phrased it, "Our great worry is how to avoid having habit take over."

One faculty member who subsequently went into business spoke of the imperative need of intimate acquaintance with a welter of details about the industry as well as the basic understanding of it which he had acquired in the university. It was necessary not only to learn the details, but also to establish relationships with the people involved in those details throughout the organization and the industry. This process required constant shifting of the scope of attention from the general to the specific, from the past to the future, from personal values to technical possibilities. In industry, the former professor was constantly involved in an artistic matching of ends and means cutting across all those subdivisions which are made more or less arbitrarily within the academic community.

Participants were left with the haunting thought that the executive role in business might be a place of more active learning than that of a professor in a school of business. Again the need was to develop both the capacity to generalize policy direction for a company and to generalize a system of knowledge under controlled conditions. There was endless discussion of the kind of curriculum which could give students and faculty

---

[1] *The Uncommon Man: The Individual in the Organization,* by Crawford H. Greenewalt (New York: McGraw-Hill Book Co., Inc., 1959), p. 64.

the capacity to learn, as well as information; and the capacity to get things done, as well as a scheme of analysis.

On the Social Structure of Academic Communities

A sixth point of emphasis was on the academic community and professional advancement for faculty and for students. During the discussion of intangible objectives of education, several participants interrupted with pragmatic reactions about what students needed in order to get jobs and what professors needed in order to be promoted.

The visible differences in teaching methods were symptoms of underlying differences in educational objectives and opportunities. The latter are most difficult to talk about because they represent a university's historical views which attach prestige to faculty membership or to degrees granted. They include the whole range of unanswerable questions: "Why do students come here?", and the similar question: "Why do professors come and stay here?" These intangible aspects of a university's prestige frequently include the access to libraries, laboratories, data in the field, and, most importantly, peers with common interests.

For faculty members, salary scale and promotion policy were obviously important factors, although not always mentioned in open meetings. Over the 11 years, 1955–1965, the spread in salaries among participants narrowed, or at least the minimum was raised substantially. The intangible aspects of university prestige in the short run are relatively fixed conditions, whereas the educational objectives and the evolution of appropriate methodology are elements in the long run which an individual professor can affect as well as be affected by. Participants enjoyed the opportunity of talking about promotion policy at the same time they discussed faculty organization and assignment of faculty members to courses in different programs. The Harvard deans each appreciated the opportunity to meet faculty members from other schools and get their reactions and questions. So much so that, on more than one occasion, it was necessary to be reminded about the original tacit agreement made with Dr. Thomas Carroll of The Ford Foundation not to pirate away the professors who had been sent to participate in this study of the case method.

Over the years another of the topics which always created a great deal of interest in the Program was the selection of students. At one extreme of participants were those from state universities which must accept any graduates from a high school in the state as freshmen in the college of commerce. At the other extreme were participants from graduate schools who accepted only a limited number of selected candidates after demonstrated performance at the bachelors' level and on the Educational Testing Service graduate business school aptitude tests. Dr. Lewis B. Ward, now at the Harvard Business School, spoke to the Program in 1955 and 1956 when he was an officer at ETS developing this test. He took pains to point out that what could be measured was only a part of the student. Furthermore, the criterion of performance in the university's educational program was in itself only a part of the criterion of success in business administration. In this way Dr. Ward endeavored to emphasize the importance of a faculty's selection process and the dangers in giving too much weight to the attractively simple statistics of a person's grade on one test or in a few courses. One professor said with pride, "Our students are going someplace and they know it." Another professor reacted, "Most of my students know it is unrealistic for them to think of what they would do as president of a company."

Participants were struck by the importance of achievement motivation discussed by Professor David McClelland when he met with the Program. As the economy develops, young men from all walks of life must find a way into management. It is not just coincidence that the shift within the United States from rural to urban and industrial life was accompanied by the expansion of collegiate schools of business administration. For thousands of young men and women who had no background in industry, the business school has bridged the gap. The lag of universities to develop schools of business administration in many developing countries is holding back economic development for lack of such a bridge into industry. Achievement motivation without such administrative training is frustrating. However, business school students without motivation also require reassessment of educational objectives and methods.

Professor Charles Orth of the Harvard Business School repeatedly aroused the group in the later programs by his research into the social structure of different sections of MBA students and the relation of this structure to the climate of learning. Although the impact of student attitude was dramatic when a professor was endeavoring to use cases, the visiting professors acknowledged by the end of the programs that such climate was important under any method. A school's emphasis upon academic disciplines, professional training, science, art, or on immediately useful knowledge obviously influences which students are attracted. The criteria in each academic community are mixed, and shift from time to time. These intangible bases for evaluation of students and faculty are related. Both grading of students and promotion of faculty members were topics of great interest in the Program. How they are related is sensed, but touchy to pursue.

Discussions of academic climate were helped by deans whose educational philosophy was sympathetic to the case method but whose faculty background was different enough to suggest important contrasts. An explanation of undergraduate as well as graduate programs helped to clarify some of the possibilities of modifying the case method pattern for different students. The argument that administration could be a truly liberalizing subject in an undergraduate educational program always brought strong reactions; the problems of a faculty dealing with undergraduates and graduates in large numbers helped to identify the way men might need to shape the case method pattern in transferring it to their home institutions. Deans coming as outside speakers helped the Program see the preoccupation of undergraduate schools with student motivation and grading, leaving little time for the research and development upon which faculty members felt their promotion depended.

The participants acknowledged the wisdom of a very close liaison between the professor and his dean in introducing the case method. The question of what significance case research and writing would be given in considering promotion and salary increases for faculty members was always troublesome. One dean observed that, as more institutions use the case method, the names of the professors preparing case material will be placed before a larger and larger audience, but stressed the desirability of "combining case research and writing with other types of research and writing so as to get complementary benefits and dual publications." The deans typically expressed the need for maintaining the requirement of productivity along with released time and funds for cases, travel, and research.

Among the participants each year there were deans or professors who could have been as eloquent as any of the "outside" speakers who were brought in to talk about the institutional context in which case-course development took place. However, the mores of the group made it difficult for any one person to take the center of the stage long enough to make a complete presentation. However, on several different occasions participants were asked to come back in later years to discuss their afterthoughts on the program in light of their institutional perspectives.

Bearing in mind the major emphases of these outside speakers, participants were able to project their thinking about the case complex in the context of their own institutions. The following section reappraises experience of alumni with case method patterns as a sequel to the Visiting Professors Case Method Program.

Involvement of Program Alumni in Emerging MBA Programs

During and after the decade of the Visiting Professors Case Method Programs, participants often found themselves in the midst of debates on basic educational policy. The participants had been selected because of their promise of leadership. Therefore, it was natural that alumni would be deeply involved in studies and experiments by their faculty groups. The introduction and expansion of programs leading to a masters degree in business administration called for major reconsideration of objectives.

When the Visiting Professors Case Method Program began in 1955, only one or two of the participating schools had MBA programs. By 1965, all but one of the 20 participants came from schools which had an MBA program or the equivalent degree for it. Such a sweeping change in American schools of business administration naturally was an important influence in the activities of alumni of the Program as they returned to their home universities. Students had mixed motivations in seeking these new MBA programs, and the faculty also had mixed objectives: pioneering research and knowledge useful to business administration; development of a professional program to equip practitioners; training specialists in different functions of business; equipping college students for vocations in business; and starting men toward careers as professors of business administration.

Many of these new MBA programs featured at least one course in which cases were predominant. How much cases were used is a relative judgment. At one university, a member of the faculty said, for instance, that 75% of the courses used cases; another member of the same faculty acknowledged that some cases might be used in 75% of the courses, but in his opinion only 15% of the courses were developed around the case method.

One of the common situations reported by alumni of the Program was the way in which students coming into an MBA program from other departments in a university or from other colleges seem to be no more mature than the majors who had come up through a departmental sequence of courses in the school of business administration. The process of maturing through which men and women go during the five or six years from college entry until the end of a graduate program in business administration is one in which there is inevitably and quite properly a large component of self-discovery as well as subject mastery. Many of the men who went through the Visiting Professors Case Method Program reported that the use of cases gave them great satisfaction because of the way in which it helped them help students in this double process of discovery.

The demand by industry for graduates of MBA programs has reinforced the encouragement given to schools of business administration by the foundations to put energy into post-graduate programs. This interest, however, has not produced a consensus on the content and concepts of a professional program. There are, for instance, alumni of the program who use cases in their undergraduate courses but give lectures on marketing theory in the MBA program of their schools. In other schools, alumni of the Program use cases in the MBA program and text books in undergraduate courses.

Introduction of an MBA program often precipitated an issue of whether the orientation of the faculty should be toward a professional school or toward graduate study. One of the alumni in a university which has a strong interest in theoretical disciplines made the significant observation that "the important thing is not to graft on a method unsuited to the content of a course and program." This point of view surely reflects a central theme of the Program. In that particular university, the extension of the business school work into a graduate program was rather clearly dominated by the concept of "advanced" graduate course work. This same professor was disturbed because his undergraduate majors were much further along than the students in his course in the MBA program. His practice was to use only an occasional case in either course and then only after the students had been exposed to concepts and relevant knowledge. In his undergraduate major course he could insist upon prerequisites of mathematics. In the MBA program students were admitted from a great variety of undergraduate majors without prerequisites that the instructor needed for what he considered a truly "advanced" course in his subject area.

Elsewhere, a more deliberate attempt has been made to make the curriculum for the MBA degree analogous to the program of other professional schools, such as law and medicine. The courses are not thought of as advanced work in areas that students were supposed to have covered in undergraduate preprofessional courses.

In metropolitan centers, the MBA programs have enrolled many men and women who are employed during the day as engineers, accountants, or salesmen. The Case Method Program alumni reported that these mature students were frequently being encouraged by their companies to gain skill in administration or were seeking personally for a way to get out of "the rut" in which they saw themselves as specialists. The use of cases in teaching such students was enthusiastically accepted by both students and faculty members, because the common desire was to move from some specialized background or general liberal arts degree in the direction of administration. Students frequently wrote cases based on their own company experiences. Some of these cases were introduced into the school's curriculum and made available elsewhere through the Intercollegiate Case Clearing House.

Courses given for practicing executives in special programs have used cases extensively. Some of the participants came to the Program because their schools were undertaking a more active program for practicing executives and were therefore interested in exploring more fully the use of cases and the development of case-courses. Utilizing these executives on the campus for cases or case leads has been a rewarding source of additional materials.

Departmental Organization and Course Development for a New MBA Program

Introduction of a new MBA program revealed a great diversity of organization within schools of business administration. Some of these are noted to illustrate the importance of context for the case-course complex. Long-established schools which had characteristically given different major undergraduate programs faced severe organizational tasks in going into graduate programs. In many schools, some of the departments had well-worked-out majors and degrees, for instance, in accounting or finance. In these departments, professors had sequences of well-integrated case-courses reflecting a considerable amount of case development. When the decision was made to introduce an MBA program, a strong and frequent tendency was for each department to look about for the courses which it could respectably advance to graduate status and include in the new curriculum. Several participants in the summer Program had been nominated by their deans because they were to be responsible for development of a new course for such MBA "nonspecialists." Other alumni who had used cases in their undergraduate courses turned to the use of lecture-discussions with students in their new MBA programs. Enrollment in some of the new MBA programs was small, 15 or 20 students, and a large proportion of the students were interested in becoming teachers. They and their instructors were looking upon the MBA as a step toward the Ph.D.

The decision to introduce an MBA program frequently involved setting up a new post of Director to focus the attention of existing departments on this new concept of an integrated program. In large well-established state university schools of business, the new program had to be related to popular masters degrees in specialized subjects under the joint jurisdiction of the graduate school dean and the heads of various departments. An alumnus of the Program reported that, even though a number of cases had been written in several different departments and many case-courses were offered at the undergraduate level, the university senate still was not sure that case writing should be taken as research to meet the criterion for faculty promotion when teaching at the graduate level.

New MBA programs were aimed at turning out much-sought-for "junior executives," who typically would go into a company training program. But it was difficult to keep faculty department interests from eroding the concept of an overall management focus. In one such program, most of the 100 MBA students came from other universities or from other schools within the university. In line with the announced objective of training men for "general management," the program originally limited students to nine hours in any one subject area. Since the Business Policy course was considered a capstone of the program and students saw it as preparation for the required oral examination, nearly all candidates enrolled in it, although it was not part of any departmental offering. When the MBA program was inaugurated, the plan was for each department to develop a special offering, but very few new courses were in fact developed. Pressure increased to allow 12 hours of work in a selected department. In the view of many department heads, this would give men enough "strength" to substitute the MBA program for the MA degree in the departmental concentration. The Director of the new MBA program felt that the basic idea of turning out "generalists" was being lost.

In another university in a metropolitan center, the school had grown very rapidly from a small department of business administration, and the MBA program had begun before departments had been established. Course offerings, therefore, had been developed with an integrated program in mind. Here cases were used in many of the MBA courses as well as in the final Policy course.

Prerequisites — Sequence

The multiple programs which students are able to put together under the requirements of such diverse institutions mean that faculty members are continuously struggling with the problem of course sequence or prerequisites. This question of prerequisites is of particular relevance in a new MBA program where students are being attracted from nearly all fields of undergraduate work. The MA or MS degree in accounting, for instance, or in marketing, finance, insurance, real estate, or transportation, typically have a well-established sequence of prerequisites. Students taking these programs are expected to go into fairly well-defined vocational positions or into teaching, where precedents have already been established. In some schools with such specialized programs alumni of the Program have been given enthusiastic support for introducing case-courses in these sequences for majors or advanced degrees. A plausible argument is advanced that many young men and women need to start into business in fairly well-defined positions. If they get the administrative point of view toward their task through studying administrative cases, they will increase their appreciation of the total enterprise to which their specialized work is contributing.

In many schools, departmental majors were given two case-courses. The first was presented fairly early in the major's sequence as a core course to expose students to a wide range of administrative problems in the area. A combined case and text book was frequently used. The same faculty major group might then offer a second course, using a case book, as a capstone for seniors or for MS degree candidates in the departmental major.

Alumni of the Program teaching case-courses at a senior level or in the second year of new MBA programs frequently were perplexed with student performance. Some students who had not had prerequisite courses in the field, but who were admitted from other colleges as exceptions, seemed to do just as well in the advanced case-course as those who had had the theoretical course in conceptual scheme and principles.

The Business Policy Course

Many Program participants reported that, at one time or another, they taught a senior or graduate level course in Business Policy. Such a course was included normally in an MBA program. Many participants came to the Program specifically for the purpose of helping their schools do a better job in organizing and teaching such a capstone course.

Casebooks in Business Policy from graduate faculties of business administration have been developed from courses which are at the end of, or which run concurrently with, several other courses using cases. In programs where the Policy course likewise is given near the end of a student's curriculum after he has studied large numbers of cases, these books seem to be highly successful. When the Policy course is given as the only one which makes extensive use of cases, there are frequent stresses.

A major purpose in the Business Policy course in many graduate school programs is to help the student, who may have dealt with 500 relatively clear-cut issue cases, to learn to diagnose a less structured situation. If such cases are given to a group with very little previous case analysis, the results may be disappointing. Participants in the Program themselves found that there was not time within the eight weeks and under the plan of rotating leadership to make much progress in a sequence of diagnostic Business Policy type cases. Preference was given, therefore, to fairly clear-cut issue cases for the case discussions in the first four to six weeks of the Program.

To develop the orientation, skill, and capacity to generalize policy directions for an enterprise, is a well-nigh impossible load for one single course. It assumes that the student's exposure to other types of generalization from other orientations, with other skills, will equip him to make the radical shift to the administrator's point of view. Other courses can contribute to the objectives of creating an administrative point of view and skill. Much of the policy-making in top corporate management is based on precedents which may or may not be recognized initially when someone in the organization has to make a decision on a fairly specific issue, often as an exception or innovation. Cases dealing with such issues from departmental levels also involve policy and may be found more readily. They also have the advantage of getting students to recognize policy-making issues at a level of activity in an organization in which they may be involved soon after graduation. Several such cases developed by participants in the Program have been contributed to the Intercollegiate Case Clearing House and have been widely used in Policy courses at the undergraduate level. In several programs the MBA degree involves an oral examination in place of a thesis, and the Business Policy course using such cases is considered the standard preparation for the examination.

The Education of Undergraduates in Business Administration

The undergraduate curriculum in most universities not having business schools gives the potential leaders in American democracy a very partial, if not greatly distorted, view of the business corporation and industrial enterprise. Quite apart from whether or not one is to work in or manage such an institution, understanding these characterizing factors in American society is part of a person's basic "literacy." When the faculties in business administration withdraw from teaching undergraduates, who is to see that this phase of modern life gets adequate attention in the undergraduates' curriculum?

The Program as it developed concentrated on collegiate schools of business. Professors in liberal arts colleges also need an opportunity to see the administrative process and the way in which this point of view enlarges the sense of participation and responsibility of students for direction of the enterprises which characterize our society. The emphasis in the United States, however, has been in the opposite direction, to shift business school faculties to graduate level courses.

In undergraduate programs, the logistical problems of finding classrooms and instructors for thousands of students have taken so much effort that alumni of the Program often had little time to innovate on educational issues. New buildings for schools of business administration have been multiplying. The semicircular amphitheater design to facilitate discussion by large groups has been used increasingly and helps in the use of cases. What size of classroom to build is still not a settled question. Some believe that the day of a 45-man class is past, and build classrooms for 100 to 350 students and conference rooms for small groups. Other buildings, in contrast, are designed purposely to make sure that no class ever has more than 45 students in it.

Many alumni of the Program have returned to their institutions and arranged for converting at least one classroom into some type of hollow square or horseshoe in which students would find it easy to exchange views about a case. The men who have innovated these rooms complain that other instructors are so eager to use them that they have difficulty in getting them for their own classes. Scheduling of classroom hours over longer days also demonstrates that professors can arrange their schedules at what would appear to be impossible hours. What expanding enrollments may yet require, and what margin they may leave for case-course development or any other research, is yet to be seen.

Under the pressure of large numbers, there is justification easily at hand for identifying an undergraduate program with the sole objective of "teaching." Several alumni of the Program pointed to the tremendous number of students and excused all the lack of case development and research on the basis that this was a "teaching institution." The textbook salesman was, in effect, the primary stimulus and source of communication. By contrast, in what would seem to be a very similarly situated institution, alumni of the Program were active in getting students to write cases and discuss them in small groups as well as in large class meetings.

Curriculum and career questions in a particular university have great influence on how the individual professor carries forward his interest in the study of administration by cases or by any other means. Fortunately, there were opportunities to have alumni of the Program return and to visit many of them on their campuses to supplement the continuing correspondence.

As the present chapter outlines some of the institutional forces which bear on the professor who wants to develop cases, one may see that questions about the case method often grow out of questions about the merits of studying business administration by any method. But, although questions continue, the development and use of cases increases. The next chapter provides objective indexes of the number of cases written and used in recording the organization and growth of the Intercollegiate Case Clearing House.

Chapter VII

An Index of Case Development

Evolution of the Intercollegiate Case Clearing House

The Intercollegiate Case Clearing House (ICH) is a facility for pooling cases. It was an outgrowth of the first two sessions of the Program, in 1955 and 1956. The participants in those Programs felt that a professor who seriously intended to develop case-courses would need more material than he could develop by himself. They realized that, if a professor had access to a wide variety of case materials, he could test his own ideas against situations observed by others, and could sequence cases and concepts to meet the needs of his own students.

From the small cooperative listing with which it began, the ICH has grown to a pool of more than 7,000 cases by 1,500 contributors from over 250 institutions. The Clearing House distributed more than 1,300,000 copies of cases in 1968-69.[1] Thus it serves as a rough index of case development. The number of cases listed and sold indicates that cases are being written and

[1] The Intercollegiate Case Clearing House has continued to grow, as indicated by total sales of $443,000 in 1968-69 covering about 1,300,000 copies of cases. Rapid growth of these intercollegiate case activities outrun the efforts at statistical analysis in this chapter, which for the most part ends with 1966.

used by more and more professors in more and more schools. A review of the development of the Clearing House provides a perspective on the use of the case method pattern, in whole or in part, in a wide variety of institutions.

The basic objectives of the Clearing House have been to improve business education: (1) by encouraging case research through the recognition provided by the listing and publication of cases; and (2) by encouraging course development through making a wide selection of cases from varied sources easily and economically available.

The First Group of Visiting Professors Recommend a Clearing House for Case Material

At the final discussion period in the summer of 1955, participants suggested that contributions to a pool of cases would be a way of keeping alert to developments in industry. The group also saw the case pool as a way of developing a professor's own field with the assistance of others, although the participants recognized that there would be differences of opinion about the relationship of case writing to publication in learned journals. A further advantage they anticipated would be the development of material characteristic of various regions in the United States. The group's written recommendations are an indication of the way in which the Intercollegiate Case Clearing House grew out of the Visiting Professors Case Method Program:

> That the experimental summary and listing of new cases done this summer be continued in order to circulate information about new cases and to find improvements in the process.
>
> That the committee confer with Mr. Towl on such methods of disseminating material during the interval until other facilities might appropriately be developed. The ultimate goal would be to establish (under appropriate auspices) a clearing house for case material.
>
> That the Deans participating as individuals in the program (Coleman of West Virginia, Dunham of Mississippi, and Searles of Toledo) consider how to report these developments in some suitable way at an early opportunity to the American Association of Collegiate Schools of Business.

A tentative summary of the cases which had been written by the group had been made during the first Program in 1955. The form used was one with which Miss Virginia Jenness and Miss Grace Lindfors, Associates in Research, had been experimenting in an analysis of the case material in the files of the Harvard Business School.

The participants in the 1956 Visiting Professors Program picked up the suggestion of the 1955 group and an energetic committee made "recommendations for the advancement of case research in collegiate schools of business," including a central case publication house. Responding to these suggestions, use was made of the cross-section of cases from the 1955 and 1956 participants. With this start and using part of the funds given by The Ford Foundation for helping case development interests in other schools, in January, 1957, the Harvard Business School published "on behalf of the intercollegiate contributors" an *Intercollegiate Bibliography, Cases in Business Administration*

This, in effect, became Volume I in a series which has continued on an annual basis. The first volume listed 276 cases from 120 contributors from 32 institutions. As the Foreword said, this was "a first step toward a clearing house for case information." This volume, with the recommendations of the 1956 committee and other materials, was included in a proposal to The Ford Foundation in March 1957 to underwrite "An Intercollegiate Clearing House for Case Materials." The proposal was submitted by Dean Stanley F. Teele from Harvard University, Graduate School of Business Administration. At that time, Dean Teele also was president of the American Association of Collegiate Schools of Business (AACSB).

On June 3, 1957, Dean Paul M. Green, University of Illinois, the new president of the AACSB, wrote to Dr. Thomas Carroll, Vice President of The Ford Foundation as follows:

> I am writing to set forth the position of the American Association of Collegiate Schools of Business in connection with the Intercollegiate Case Clearing House, which has been established at the Harvard Graduate School of Business Administration. It is our understanding that The Ford Foundation is willing to support the Intercollegiate Case Clearing House to the extent of $50,000 a year for three years and that the Foundation would be recep-

> tive to a proposal from the Association that this grant be made to the Association with the understanding that the Association will in turn arrange with the Harvard Graduate School of Business Administration to carry out the assignment. This letter constitutes the proposal.

On the first of July a grant of $120,000 was made along the lines of Dean Green's proposal "on a three-year trial basis."

Under the guidance of its director, Mr. Harvey P. Bishop, the Intercollegiate Case Clearing House expanded greatly and achieved wide participation during the first three years of its existence. It did not in that time become self-supporting, but as the end of the period covered by the original Ford grant approached, the AACSB Advisory Board urged the Clearing House staff to project continued growth in keeping with the imperative need to improve education for business administration throughout the world. On October 3, 1960, The Ford Foundation wrote to Dean Maurice W. Lee, then president of the American Association of Collegiate Schools of Business, advising of an additional grant of $120,000:

> It is our understanding that the grant funds will be used by the Intercollegiate Case Clearing House to expand its operations for handling case materials in business administration and to develop measures of case selection and other procedures intended to enable the Intercollegiate Case Clearing House to become self-supporting.

This goal was achieved by the end of 1965. And before he retired in 1966, Mr. Bishop had managed to create a comfortable operating balance: a tribute to careful handling of millions of pages of case material at fractional cent margins. The Clearing House, as any growing concern, must still husband its resources against the fluctuations in popularity of cases. (See Exhibits VII-1 and VII-2 for a summary of the bibliographies available. In addition, the Clearing House has available listings of cases involving urban problems, minority groups, and other contemporary issues, as well as cases on planning, programming, and budgeting.)

Although the majority of cases submitted so far have been from the United States, increasing numbers of cases are being received from other parts of the world. There has also been a

EXHIBIT VII-1

INTERCOLLEGIATE CASE CLEARING HOUSE

SUMMARY OF BIBLIOGRAPHIES PUBLISHED, CASES RECEIVED AND DISTRIBUTED

Year Ending July 1	*1957*	*1958*	*1959*	*1960*	*1961*	*1962*	*1963*		*1964*	*1965*	*1966*
Intercollegiate Bibliography Volume No.	I	II	III	IV[a]	V	VI	VII	VIII[b]		IX	X
Cases Contributed											
Institutions	32	38	34	63	46	60	57	30		73	61
Contributors	120	151	167	304	212	270	262	106		305	297
Cases	276	389	370	710	383	550	576	165		745	587
Cases Distributed											
Sales (in thousands of dollars)	—	$ 3	$32	$47	$79	$ 75	$107		$154	$153	$214
Case Books Listed from Publishers	—	48	56	75	88	113	125	—	—	200	223

[a] In addition, Volume IV carried an International Section listing 242 cases from 71 contributors in 20 participating institutions.

[b] In addition, Volume VIII repeated 295 cases from Volumes V and VI which had been most widely useful.

EXHIBIT VII-2

BIBLIOGRAPHIES AVAILABLE FROM
INTERCOLLEGIATE CASE CLEARING HOUSE

Cases in Business Administration:
- Selected Cases, I-X, Indexed Volume, 1967
- Preliminary and Partial Listing, 1968
- Cases in Data Processing, 1965
- Cases in Research and Development, 1965
- Cases in Technological Change and Automation, 1965

Cases in Administration of Higher Education, 1966

Cases and Other Materials for the Teaching of Multinational Business, 1964

Cases and Other Materials for the Teaching of Business Administration in Developing Countries:
- Latin America, 1966
- South and Southeast Asia, 1968
- Africa and the Middle East, 1969
- Comparative Index, 1969

very gratifying increase in classroom orders for Clearing House cases. More than 1,300,000 copies of cases are being distributed annually to more than 500 educational programs. A professor contributing a new case to ICH now has it called to the attention of more than 4,000 other faculty members, in collegiate schools of business, who have expressed interest in learning about new case development. Thus, the major extension of case development and case use is clear. However, the meaning of these ICH growth figures requires further analysis for our present purpose, which is to understand the context of the case method in different institutions.

The major means by which most students get cases is a printed case book from a commercial publisher. In 1966 there were 223 of these printed books listed in Volume X of the Intercollegiate Bibliography (see Exhibit VII-3). The increase in number available, as compared with the 48 listed in Volume II in 1958, indicates how private book publishers estimate the increasing demand for case material.

The limits to the market for case books can be measured to some extent by the fact that there were 130 member schools in the AACSB in 1968. Estimates run to something like a total of 600 institutions in the United States that give degrees in business administration including departments or divisions of business administration as well as schools. Thus, adoption of a case book in 200, 300, or more programs indicates exceptional popularity. Adoptions by 100 departments are not infrequent, but case book publishers indicate that 25 to 75 is more common and many case books continue in print with fewer adoptions. The number of students in the courses in which a case book has been adopted may vary from 5 to 1,000 or more. The population explosion promises to increase the demand for all kinds of teaching materials.

EXHIBIT VII-3

COMMERCIALLY PUBLISHED CASE BOOKS
LISTED IN THE
INTERCOLLEGIATE CASE CLEARING HOUSE

	Volume X 1966	*Volume II 1958*
Marketing (Mkt. Res., Adv., Sales Mgmt., Procurement, Wholesale & Retail)	42	12
Business Policy, New & Small Enterprise, General Mgmt.	32	5
Production Management, R & D	24	5
Human Relations	24	6
Finance	19	4
Personnel & Labor Relations	19	9
Accounting, Managerial Acct., Controllership	18	3
Government & Business, Business Law	18	2
Managerial Economics, Statistics, Operations Research	17	1
Transportation, International Bus., Real Estate, Life Insurance, etc.	10	1
Totals	223	48

Against these measures of mass distribution, the numbers of copies of individual cases sold by the Clearing House are small, clearly meeting the limited specific needs of particular course objectives. The Clearing House must cover its expenses from orders which, for example, typically run from 25 to 75 copies of from ten to a dozen cases. The few large volume orders must also offset the cost of samples which are sent out by the thousands. However, supplying samples to help professors design a sequence of cases and concepts for the needs of a particular class of students is a major function of the Clearing House. It is hoped that enough of these small custom-made courses will grow and result in large enough orders to cover the costs of additional creative efforts as the process of innovation continues.

The scale at which ICH operates is further illustrated by the distribution of orders for a currently popular case in 1965, as follows:

No. of copies	*No. of orders*
less than 25	4
25 and less than 50	8
50 " " " 75	5
75 " " " 100	5
100 and over	2

In total, 1,500 copies of this case were distributed in 1965.

Another indication of sales characteristics of cases in ICH is the sales summary of cases written by the 227 participants while attending the Case Method Program:

Cases of 157 participants had average sales of less than 50 copies a year
Cases of 50 participants had average sales of 50 to 199 copies a year
Cases of 20 participants had average sales of more than 200 copies a year

Only 6% of these cases had total sales as large as 1,500 over a five-year to seven-year period. Furthermore, the number of copies sold fluctuated, because many cases were ordered by only three to ten schools or executive programs, as illustrated in Exhibit VII-4.

About two-thirds of the cases originally listed were of such specialized interest that after two or three years of sporadic orders they were no longer carried in stock. This was not an unexpected rate of attrition. The survival for ten years of even one-third of the cases testifies to the tremendous range of course and student needs around the world. The Intercollegiate Case

EXHIBIT VII-4

NUMBER OF CASES SOLD BY
INTERCOLLEGIATE CASE CLEARING HOUSE

	1959	*1960*	*1961*	*1962*	*1963*	*1964*	*1965*	*Total Number of Copies Sold*
Case W	35	325	430	100	400	530	80	1,900
Case X	250	190	120	280	180	284	55	1,359
Case Y	—	—	—	70	350	192	210	822
Case Z	—	75	100	50	570	65	75	935

Clearing House fills its role in part by helping four or five professors here and there to enrich their courses by making available a particular case in which no one else seems interested. For the economic survival of ICH, it is hoped that there also will continue to be a number of cases which meet the needs of a thousand students year after year, to provide the economy of long, low unit cost printing runs.

Since the 227 participants in the visiting professors programs represent a cross-section of subject interests as well as institutional backgrounds, the experience of ICH with sales of their cases provides perspective on the relatively few cases which meet the needs of large numbers of students, and the relatively large number of cases which continue to meet the particular needs of only a few professors and their students.

The schools which have contributed most of the cases and those which use most of the cases do not necessarily overlap. Exhibit VII-5 lists in alphabetical sequence the schools that have contributed the largest number of cases held in stock by ICH in 1966, and the schools that were the largest purchasers of cases from ICH in the three years: 1964, 1965, and 1966. From this compilation, it is evident that the case method complex has been instituted in some measure in a wide variety of institutions. Happily there are a substantial number of additional schools contributing and using cases. Furthermore, the list does not show several institutes outside the United States where a special case development effort has been made. A separately financed series

EXHIBIT VII-5

INTERCOLLEGIATE CASE CLEARING HOUSE
RELATIONSHIPS WITH EDUCATIONAL INSTITUTIONS
Contributors and Purchasers

Educational Institutions	*Educational Institutions*
Alabama	Michigan
Alberta	Michigan State
Arkansas	Missouri
Bentley	New Brunswick
British Columbia	North Carolina
California (Berkeley)[a]	Northeastern[a]
Chicago	Northwestern[a]
Columbia[a]	Oklahoma
Colorado	Oregon
Connecticut	Pennsylvania (Wharton)
Cornell	C. W. Post College
Dartmouth	Purdue
Fresno State	Rensselaer Polytechnic[a]
General Motors Institute	Rutgers[a]
Georgia	South Dakota
Harvard[a]	Southern California
Hawaii	Stanford (incl. ICAME)[a]
IMEDE[a]	Texas
Indiana	Texas Christian
Kansas[a]	Virginia (Graduate)[a]
Kentucky	Washington (St. Louis)
Louisiana State	Washington (Seattle)
	Western Ontario[a]

[a] Schools with 10 or more cases still in stock and purchases of more than $2,000 in 3 years, 1964-66.

of bibliographies for teaching business administration in developing countries has been prepared.

For the 10 volumes of the Intercollegiate Bibliographies, 188 institutions have contributed a total of nearly 5,000 cases in the following proportions:

135	institutions contributed	1-10 cases
41	" "	11-50 "
12	" "	51 or more cases
188		

In other words, one or a few professors in many schools have developed some case material, but the majority of cases come from a relatively few schools. The Program and subsequent reactions from "alumni" have pointed out the significance to the professors and institutions of even a relatively small amount of case development. About two-thirds of the Program participants are working (in 1968) in schools which have developed less than 10 new cases during the 10 years of ICH. However, the fact that they understand case research and writing enriches the use they make of printed case books and loose-leaf material secured from ICH. The distribution of participants in the Program by the cases contributed to ICH from their institutions was as follows:

Number of Cases Contributed to ICH	*Number of Schools Contributing Cases*	*Number of Participants in the Case Method Programs*
Less than 10	85	150
10 and less than 20	19	45
20 or more	12	32
	116	227

About one-half of the Program participants have sent in additional cases in subsequent years. Thus, even though they may have received relatively little institutional support, they have found ways to get released time, to get student help, to convert consulting reports, or to utilize executive development program contacts. The resulting case obviously does not constitute a course, but active creation of even one or two cases vitalizes the collection and sequencing of other materials. The leverage which a professor gets from creating even one section of a course along the lines of his own conceptual scheme helps to lift the burden of class after class, and thus contributes to the professor's own professional development. This follow-through of participants was discussed more fully in Chapter 6.

Finally, the intrinsic value of the cases developed during the Program and all those listed in ICH must not be minimized. These cases represent significant building blocks in the structure of better education for business administration.

POLICY DIRECTION OF THE INTERCOLLEGIATE CASE CLEARING HOUSE

The Advisory Board for ICH is the Executive Committee of the American Association of Collegiate Schools of Business, the president of AACSB being ex officio chairman of the Advisory Board. The list of chairmen (see Exhibit VII-6) indicates the broad range of institutional interests which helps to make ICH responsive to the needs of all kinds of institutions. Members of the Advisory Board have brought to the policy questions of ICH in the past decade the experiences and points of view of 40 different schools. Many of these deans sent one or more of their faculty members to the Program and some have participated themselves.

EXHIBIT VII-6

INTERCOLLEGIATE CASE CLEARING HOUSE
CHAIRMEN OF ADVISORY BOARD, EX OFFICIO

(Presidents of American Association of Collegiate Schools of Business)

1957/58 Paul M. Green, University of Illinois
1958/59 Sidney G. Winter, State University of Iowa
1959/60 Ross M. Trump, Washington University
1960/61 Maurice W. Lee, University of North Carolina
1961/62 Richard Donham, Northwestern University
1962/63 Austin Grimshaw, University of Washington
1963/64 R. F. Patterson, University of South Dakota
1964/65 Paul Garner, University of Alabama
1965/66 Charles J. Dirksen, University of Santa Clara
1966/67 Willis J. Winn, University of Pennsylvania
1967/68 Donald J. Hart, University of Florida
1968/69 Floyd A. Bond, The University of Michigan
1969/70 Paul V. Grambsch, University of Minnesota

A brief review of some of the policy issues discussed with the Advisory Board will help to give perspective on the case method in a variety of schools. At the first meeting of the Board on October 22-23, 1957, there was prolonged discussion of the issues which the new intercollegiate facility for case reproduction and distribution raised. While the Advisory Board has urged the staff to experiment and many changes have been made from time to time, the statement of fundamentals developed at that first meeting has served virtually without change. These fundamentals have been reproduced in various volumes of the Intercollegiate Bibliographies and on the reverse side of information sheets submitted with each new case to be listed. On the form dated Fall 1966, they were as follows:

> In the development of the Intercollegiate Bibliography and Case Clearing House the following fundamentals are stressed:
>
> *Maintenance of Business Confidence*
>
> The continued success of case research and distribution depends upon the confidence of the business community; each professor, therefore, shall be responsible for professional treatment of experiences recorded by him and for securing proper authorization from the company for educational use.
>
> *Recognition of Case Research*
>
> The creative task of discovering the significant facts in an administrative situation and summarizing them in the form of a case is recognized as a significant professional contribution which the Advisory Board shall seek to encourage by developing facilities to circulate the finished cases among professional colleagues.
>
> *Encouragement of Course Development*
>
> The administration of the Intercollegiate Bibliography and Case Clearing House shall also seek ways to encourage professors in the selection of cases and their arrangement in courses to meet the needs of specific student groups.
>
> *Ownership and Control of Cases*
>
> The contributing author or institution shall retain all proprietary rights in the cases, including reproduction for classes and inclusion of such cases in printed case books, but shall grant to the Intercollegiate Case Clearing House the right to list, reproduce, and sell copies of the cases contributed without royalty. The contributing author or institution should make any copyright arrangements and insert the copyright notice desired on the first page of any case submitted.
>
> *Cooperation with Commercial Publishers*
>
> While encouraging the exchange of new cases for new courses, the Advisory Board also recognizes the values of a printed collec-

tion of cases for well established courses and shall encourage those professors who are able and so disposed to develop books of their own cases for commercial publication.

When an author who has contributed a number of cases decides to develop a case book of his own to be distributed by a commercial publisher, he may inform the Intercollegiate Case Clearing House of the title of his book, the publisher, the date of publication, and which of his own Clearing House cases are included. The Intercollegiate Case Clearing House will then put this information on the cases that he has contributed to the book and thereafter will accept classroom orders from one instructor for no more than approximately five of these cases to supplement other cases and materials in a course which the instructor is developing.

The Intercollegiate Bibliography shall list the printed case books currently available through commercial publishers.

Concern for Commercial Case Book Publication

The first policy issue actually arose even before the Clearing House came into being. As the negotiations for the financing of the Clearing House through the AACSB progressed, Dr. Lloyd G. Reynolds, Director of the Program in Economic Development and Administration of The Ford Foundation, wrote to Stanley F. Teele (then Dean of the Harvard Business School), expressing the concern which had been raised within The Ford Foundation about the possibility that using tax-free funds might appear to undermine private enterprise in the form of a book publishing industry. The Clearing House staff discussed this problem with several commercial case book publishers, who made it clear that ICH would not be competitive but complementary to the work they were doing. The expectation at the founding of the Clearing House that the two activities would be complementary is borne out by the increase in the number of printed case books (already tabulated in Exhibit VII-3) as the activity of the Clearing House involved more professors and institutions in the case method complex.

In 1957 the Advisory Board realized that, since most professors teach several courses, they usually have time for only a token amount of active case-course development. When a professor generates a few cases of his own and exchanges them with other professors of kindred interests through the Clearing House, he feels rewarded by his creative achievement. Even though he himself does not have the time to write more cases, his interest in case books, commercially available for use in other courses, is increased. The expansion of both the number of commercial case books and the number of cases contributed to ICH during the decade bears out this original concept.

The Advisory Board also was aware that the heavy costs of the field research and basic case preparation were typically not charged to the case book publisher. It therefore seemed proper to proceed with experiments in a clearing house which would encourage more widespread case and course development and, at the same time, encourage professors "to develop books *of their own cases* [author's italics] for commercial publication." In such instances there has been little difficulty in working out the limitation on the ICH distribution of cases. However, there have been operating questions, for instance, as to how ICH was to know when cases were included in a book; or how a series of cases from one company should be counted; or the effect of a printed anthology of cases on a given subject drawn from a number of contributors to ICH. The expanding student population makes a collection of cases look like an easy and profitable answer to the demand for publication. However, as detailed in Chapter IV on case-course development, a soundly conceived sequence of cases tested in the field and in the classroom is not easy to create.

Many new possibilities appear as professors experiment with cases available from the Clearing House in relation to printed case books or other course materials. For instance, there is a trend to refer to ICH cases in teachers' guides to commercial case books as supplementary case material or for use in examinations. A new development is the reference to Clearing House cases by the authors of textbooks that do not contain cases. For instance, the Instructor's Manual to accompany the second edition, 1962, of *Financial Management* by Professor Robert Johnson, Michigan State University, refers to 6 commercial case books and in addition specifically mentions nearly 200 cases to be used in connection with the various chapters of this textbook. Another example is an *Introductory Analysis of Marketing* by

Professors Matthews, Buzzell, and Levitt of the Harvard Business School and Professor Frank of Stanford. There are no cases in this book, but it has an Appendix which refers to four commercial case books and, in addition, to many cases taken from the Intercollegiate Clearing House to be used with individual chapters of the text.

There are a number of implications which merit further study by individual researchers, educational institutions, and commercial case book publishers, as well as the Intercollegiate Case Clearing House. New opportunities appear to be emerging to get maximum distribution of case materials for the use of students in a wide variety of course arrangements while still encouraging individual authors by proper recognition. Professors who have been unable to develop enough cases for a book of their own appear to be pleased to have their cases included with others to round out and give variety to cases in popular case books.

ICH has listed large numbers of cases for which there is limited demand, and has listed other cases which became very popular, but for which there was a relatively short economic life before their distribution was limited by their inclusion in a commercially distributed printed case book. The record since 1957 shows clearly that ICH has filled a niche in case-course development around the world which augments rather than constrains the distribution of cases by commercial publishers.

Once a sound basis for this relationship between the Clearing House and commercial publishers had been established, the Advisory Board turned its attention to other questions.

Editorial Policy

How does anyone tell whether a case is good or poor? Should the Intercollegiate Case Clearing House accept every case submitted? The Advisory Board continued to urge (through the evolution of ICH) that the staff select high-quality cases, but exploration of feasible ways to determine quality led more and more to reliance upon "survival of the fittest." Early discussions with the Advisory Board established some outside limits, and the requirements of becoming self-supporting hastened others.

In the first Advisory Board meeting, lengthy discussion of editorial policy grew out of the thought that the Clearing House should take an active editorial position and refuse poor cases, but the Board recognized the difficulties of getting an editorial committee to pass on each case. In addition, it became clear in reference to some specific case experiences with members of the Advisory Board that what might be a very "thin" case from one point of view might be an excellent starting point for discussion with another group of students. The varying needs of undergraduates, specialists, and executive programs, for instance, indicate why the issue has persisted throughout the life of the Clearing House. Agreement was easily reached that the Clearing House should concentrate on making available direct field experience with administrative situations. The signature of the author or his certificate that the case had been authorized for reproduction and for widespread distribution by a responsible party in the situation became a clear criterion.

Another of the topics which required considerable discussion was the treatment of cases prepared in companies or by government agencies. If the Clearing House were to operate on the basis of accepting field cases without selection, a consistent policy for dealing with contributions of cases from companies which might verge on public relations or advertising stories seemed imperative. The Advisory Board recognized the valuable aid to education which many corporations were providing in the form of cases developed for their own internal training programs. Since many of these programs involved faculty members of different universities, it was agreed that such company-originated cases should be accepted only when used by and sent in by a faculty member of some university.

The occasional receipt of an extremely long case or a case with complex exhibits led the staff to send such cases to men in other schools known to have an interest in this type of material. These readers generally replied that the material, however rich it might be in empirical data, did not promise to meet classroom discussion needs. The Advisory Board agreed that research monographs about case situations were off the main ICH target, which is to involve students in analysis and action in the classroom.

The Clearing House also received numerous cases written by students under the supervision of faculty members. While recognizing the value to students of this field research and case writing, the Advisory Board and staff have felt that the professor should take responsibility for selecting and testing such cases before endorsing them for use by others.

The Advisory Board also encouraged the development of an international supplement, but postponed the handling of cases in languages other than English. The special bibliographies for developing countries has involved experimentation with cases in other languages.

The perennial question: "Should we seek to get cases listed as quickly as possible or do so only after they have been tried in class?" still requires examination of the logistical features of production and distribution of cases in looseleaf form.

The Logistical Problems of Looseleaf Case-Course Materials

Most university communities have bookstores. Until they become organized to handle looseleaf material, an alternate central distribution point is frequently an important step in making case-course development feasible. Some states specifically forbid faculty members to take money from students for materials. Thus, ingenuity is required to provide a case laboratory or some other channel for getting teaching materials to the students without involving faculty members or the university itself in the handling of funds.

When the Advisory Board met on February 17, 1959, there were several of these practical matters among the 16 questions which the ICH staff listed for discussion. The use of cases introduces some logistical problems which the lecture avoids in its historically economical means of communicating knowledge with a minimum of materials for use by students. The printed case book puts into the hands of students a fixed package of material for which there often is an established resale market. The printed book, however, limits the instructor to the conceptual framework of the author. The distribution of looseleaf materials which the professor has sequenced according to his own course concepts and according to his estimate of the needs of his particular students does require a somewhat different type of distribution system. Further studies must be made of the total logistics of getting cases into the hands of students. The collating service provided by the Intercollegiate Case Clearing House for putting packets of cases together has helped the college bookstores which are set up to handle only bound volumes from established commercial sources. The Clearing House also has been able to provide a limited number of free sample cases for inspection.

In spite of the general increase in prices of almost everything else, the prices suggested in the first Advisory Board meeting of 5¢ a page for training programs in business firms and 1¾¢ a page for students in nonprofit educational institutions were not instituted. Prices for training programs in business firms were set and kept at 3¢ a page and at 1½¢ a page for students in nonprofit educational institutions until January 1, 1968. Prices were then raised to 4¢ a page with a 50% discount for university programs. Evolution of pricing policy is further discussed under operating requirements of ICH.

The Contribution of a Teaching Note

Agreement on the twin fundamentals of "Recognition of Case Research" and "Encouragement of Course Development" left the staff and Advisory Board with additional operating problems. Although there is some duplication between those professors who contribute cases to ICH and those whose orders provide the operating funds for it, there are also some differences. Even the same man seems to be thinking differently when submitting a new case and when looking for additional cases to fill places in his course design. It is hard for him to describe the new case in a way which will help another professor tell whether or not it will fill a need in his course. Even some professors of advertising do not realize that the desirable qualities of their new cases need to be described in the bibliography in such a way that others may know what is available.

The development of the Intercollegiate Case Clearing House presents again the different emphases given to "the case method" by different professors. At one extreme are professors who tend to think of the case method as a method of instruction — an alternative to a lecture or a chapter in a textbook. At the other extreme are professors for whom the case method is a method of studying administrative situations in the field.

With the instructor's needs in mind, the desirability of a case appearing in the bibliography is clearly enhanced if a teacher's guide also is listed. In the Program, professors were encouraged to prepare a teaching note as well as the case itself. This device helped to make sure that the case included the information that would be needed in the classroom. It also helped instructors see the point of view from which the case researcher was selecting data. It would be unfortunate abuse of such guides if they were looked upon as the basis for a lecture at the end of the class in which the professor's superior wisdom could be demonstrated. Short of that danger point, the utility of the case is clearly enhanced if the context for its use is sketched out.

The importance of such clues in the abstract of the case may be better understood by considering the Intercollegiate Case Clearing House as something of a common carrier. There are, indeed, many different documents which could legitimately be called cases. The administration of the Clearing House has felt that it should accept whatever a professor considered to be a case, with a few gross exceptions. The test of usefulness by a significant number of professors and students is the pragmatic criterion. From the Clearing House, therefore, a wide range of cases is being used by a wide range of professors with a wide range of both graduate and undergraduate students, and by executive development programs.

A great variety of cases are listed. Cases are available, for instance, to allow "decision making" by "the simplex method: a tool for decision making." Here in a concrete set of data there is opportunity to find the optimum means to an expressed end. The Clearing House also has a large number of cases in which the ends are not clear. In them, students need a different type of decision making, situational thinking, to determine direction, purpose, goals. In both types of cases there is need to distinguish between the act of choice and the process of introducing new routines into the patterns of behavior in the organization.

Anyone leafing hurriedly through the abstracts of the 7,000 cases in the Intercollegiate Case Clearing House cannot help but be impressed with the tremendous variety of situations which faculty members and students around the world are struggling to understand and for which they are trying to design appropriate action. The Intercollegiate Case Clearing House has been an important facility for allowing alumni of the Program and colleagues from other schools of business administration to exchange the results of their case research.

Efforts now under way to build up indexes, and to make case materials in this growing pool more widely and readily available, emphasize the need for continued communication. The vigor of developments in divergent directions must not be restrained by imposing rigid nomenclature. The process of successive approximations needs to be accelerated. Indexes must call attention to materials which look alike but reflect basically different interests. Such distinctions may help to accelerate the division of labor within the academic community while permitting the reassembly of the results by the students.

Operating Requirements of ICH

The first grant from The Ford Foundation permitted the Advisory Board and Staff to introduce a major change in policy on distributing samples and filling quantity orders. When the first bibliography was published, prior to this grant, there was no central stock of cases. Anyone wishing a sample had to request it from the author and get his permission to reproduce copies for class. By the time of the October 1957 meeting with the Advisory Board, a large number of reactions had been received from professors all around the country. These men told of frustrating efforts to get copies of cases which appealed to them when they read the bibliography. The author of the case, however, would often reply, if he replied at all, that he did not have copies available. Since it was to help meet this situation that The Ford Foundation made its grant to the AACSB for the Clearing House, the Advisory Board decided that the Clearing House staff should have copies of all cases on hand from which to fill requests for samples and volume orders and take a firm position against reproduction by others.

At first, the policy was to provide each contributor with an initial run of 35 copies of his own case. This reflected the prevailing situation at that time when many schools were on such limited budgets that a professor was not able to get copies of his own new case duplicated in his home institution. The Clearing House, therefore, provided the 35 copies which would allow most contributors to try their new cases with their own students.

After three years of operation, when the Advisory Board met on February 9, 1960, the interest among schools of business had increased, as evidenced both by an increase in the number of contributors each year and by the number of courses using cases from the Clearing House. The predicament of the Clearing House was similar to that in many small, growing enterprises. Expansion of sales put working capital under strain. The rate of growth by ICH could have been limited to the working capital available, but that would have limited other developments in education for business administration. With the advice of the deans of the AACSB and the support of The Ford Foundation, plans were made to expand the services and to cover deficits by a further grant. Thus it was possible for the Clearing House to continue to send authors 35 free copies for classroom use and to make sample copies liberally available to faculty members contemplating the introduction of cases into their courses. The possibility of discounts on quantity orders was considered, but the number of schools ordering 500 or more copies of individual cases was so small that no settled policy was deemed necessary. Under all these circumstances, it then seemed advantageous to list cases as quickly as possible to encourage contributors and to make a wider cross-section of cases available for all types of students and courses, even though the attrition was higher than if cases had been tried out more thoroughly in class and revised before being sent into the Clearing House.

The additional grant from The Ford Foundation made it possible to hold down the prices and thus further encourage case use.

By the summer of 1960 the economics of the Intercollegiate Case Clearing House seemed well-established, and the direction in which it should move to become self-supporting seemed evident. Balancing the budget required a favorable ratio of high-volume, low-unit-cost, established cases to small-volume, high-unit-cost, new cases. Listing the latter, however, as a service to new case researchers, would continue to be a major objective. On the advice of the Advisory Board, the price announced in Volume V of the Intercollegiate Bibliography issued in December 1960 was increased to 2¢ a page, and a limit was put on free samples. But the effectiveness of other changes by the time the volume was published made it unnecessary to implement the price increase, until January 1968.

The Advisory Board agreed that the staff should require professors contributing cases to the Clearing House to send in at least 10 copies which could be sent out as samples until requests for the case justified the expense of reproduction. This reversed the original practice of giving contributors 35 free copies of their own cases. Concurrently with expansion of the Clearing House, more schools were able to duplicate their professors' new cases so that they could be tried and revised before being sent to ICH. The listing of such tested cases seems to have been more useful, and surely more economical, than the effort to distribute untried new cases immediately.

To encourage larger volume sales of cases with low unit costs, there were listed in the fifth bibliography of the Clearing House in each section the cases from previous volumes which had been most widely used. And a request was made to users of case material to send in a list of those cases which they would most like to see the Clearing House continue to keep in stock. Otherwise the list would be trimmed back to a self-supporting number of cases with large sales. Thus experiments began to determine ways of increasing sales volume of selected cases to compensate for the cost of making new cases more quickly and easily available.

In addition, a careful review of sales was made and a special bibliography of *Selected Cases* was published in May 1961. In essence, the cases were chosen by the users of the cases, because selection was on the basis of the number of copies sold and the number of orders received from various sources. Out of more than 2,000 cases which had been submitted, about one-third thus promised to be popular enough to reprint and stock in large volume. In 1963 another selection of the most popular third of the additional new cases was published. Regardless of whether or not the most widely used cases are necessarily the best cases, emphasis on them has been essential to the continuation of the Intercollegiate Case Clearing House on a self-sustaining basis.

While concentration upon widely useful cases enabled the Clearing House to exceed the 1960 five-year sales projection by 50% in 1965, this does not mean that other cases submitted were poor. Some situations are of significance once and then are superseded. And, as in all research, progress comes by trial and error. Furthermore, new course developments provide a context for cases which were at first passed by. These reasons and others caution against extending inferences about the quality of cases without careful analysis. The Clearing House appreciates the contribution of popular cases to its own survival, but seeks to index all contributions for the encouragement of the constant probing and experimentation by which the profession of business administration will be enriched.

Evolution of the Intercollegiate Case Clearing House over the past decade into a self-supporting facility for the exchange of cases helps to answer the question about "the transferability" of the case method. The method is being used in a wide range of institutions. The pattern of use varies from school to school. It also varies from time to time in the progress of an individual professor's career. Much is being done in many places even without all the supports which were available in the Program. The commercial case book publishers and the Clearing House give professors great resources for developing their own courses for the classroom.

Much progress has been made toward the objective of making cases readily available for creating new courses. Slower progress is being made toward the objective of encouraging professors to get out in the field, to collect primary data, and to test their course concepts for themselves.

Chapter VIII

An Innovation in Cooperative Case-Course Development: The Southern Case Writers Association

Few schools had smaller resources to use for developing new cases in 1955 than those in the southern states. Yet a decade later the participants in the Program from southern schools had far more than the average to show for their continued case-course development. These participants also were very much involved in a pioneering effort to relate cases in business policy to developments in the behavioral sciences. How this came about is the account of the organization and activities of the Southern Case Writers Association. The experiences of this group of professors touch on many of the institutional problems facing those who would like to go further in case-course development.

In the first program in 1955, three deans participated: Ray Coleman of West Virginia, Clive Dunham of Mississippi, and Clair Searles of Toledo. These three were asked by all their fellow participants to explain the Program to the American Association of Collegiate Schools of Business. They not only carried out this commission for the whole AACSB, but also especially impressed the deans of the other southern schools of business administration. Nominations from the south increased markedly. But at that time the southern schools of business administration typically had very limited budgets; uniformly, the regret of participants upon their return was that they were unable to get the time or the funds to carry on further case research. For instance, the salary figures of professors, collected by these alumni in preparing their case research proposal to The Ford Foundation, ranged from $6,500 to $9,800 for 9 to 10 months, with at least a 12-hour teaching load.

The participants from the south did meet each other in the fall of each year during sessions of the Southern Economic Association, which alternated between a location such as Atlanta in the east and Memphis in the central south. Many of the participants also saw each other in the spring of the year when the Southwest Social Science Council met. Taking advantage of these repeated gatherings, in the fall of 1957 when the professors met in Memphis, Professor William T. Hicks from Mississippi organized a discussion on preparation of business cases and a reunion of Program alumni. At this meeting, the common need for support, and the recognition that cases with more southern characteristics were needed, led to the establishment of the Southern Case Writers Association. M. M. Hargrove, Dean of the College of Business Administration at Tulsa, was selected as president. There was some ambiguity then whether this was to be an association of the schools who had sent men to the Program or an association of the men who had been there. Some of the schools were concerned that participation in a grant to a regional group might prejudice their chance of getting a substantial direct grant. Ultimately when funds were in hand, an ingenious practice emerged to keep membership among individuals but to involve their deans in agreeing to release time and to provide local expenses. Generally, funds were disbursed by the Southern Regional Education Board as fiscal agent via the participant's own institution.

In the spring of 1958, Dean Hargrove explored with The Ford

Foundation the possibility of support for a regional case research effort. Dr. Neil Chamberlain of The Ford Foundation met with Dean Hargrove and others at the Southern Social Science Council meeting later that spring.

Meanwhile for the summer program in 1958 there were an impressive number of excellent nominations from southern schools, evidencing the widespread desire to build stronger and more effective faculties of business administration. There were so many nominations from the south that the geographical cross-section that year was stretched to take half of the Program from the south, 10 men. Thus, when the Southern Economic Association met in Atlanta in November 1958, there were 23 Program alumni who came for the conference called by Dean Hargrove as president of the new Southern Case Writers Association.

The cases written by these men from the south while in the Program, along with a few cases which had been submitted to the Intercollegiate Case Clearing House by others from the south, were pulled together for a special presentation at this Atlanta meeting of the Southern Economic Association. Listed were:

5	cases in Control
15	cases in Finance
10	cases in Human Relations
6	cases in Marketing
4	cases in Production
9	cases in General Management
—	
49	Total

There was a widely felt need for more business policy course cases which reflected conditions in the region.

The common experience that most of these men had had in the intensive case method program gave them a foundation upon which mutual interests soon developed. Nevertheless, in these first meetings at Memphis and Atlanta, much time was taken in merely getting acquainted with other participants in the Program and their guests. In Atlanta, an Executive Committee also was elected, including M. M. Hargrove (Tulsa), Chairman; Curtis E. Tate (Georgia), Secretary; Preston Le Breton (Louisiana State); William T. Hicks (Mississippi); and Ralph Thompson (Florida).

The Executive Committee was authorized to submit a proposal to The Ford Foundation to cover the costs of released time so that members of the Association might prepare about 60 general policy cases over the next 3 years. Funds were to be received and disbursed through the Southern Regional Education Board in Atlanta. An interim meeting of the Association was set for the spring of 1959.

In the interval, the Executive Committee had significant interchanges with officers of The Ford Foundation, and Dr. Neil Chamberlain, Director of the Program in Economic Development and Administration, agreed to meet with the entire group.

Concern with Organizational Theory Added to Business Policy

"Unless it is feasible to generalize from experience, the study of concrete situations has no meaning." Thus Neil Chamberlain of The Ford Foundation reacted to the proposal which had emerged from the Atlanta meeting. Later when he met with the Southern Case Writers Association in their spring meeting of 1959 at Galveston, Texas, he expressed the thought that case materials should be developed to substantiate or refute some of the existing organizational theories formulated in various disciplines.

Also present at this meeting as social science consultants were Harold J. Leariol of Carnegie Institute of Technology and Raymond Bauer of Harvard University in addition to 13 alumni[1] of the Program and four others from a total of 11 southern schools of business. In their discussion of Dr. Chamberlain's point of view, the visiting social scientists and others acknowl-

[1] By 1962 the Southern Case Writers Association included 58 professors from 20 universities.

edged important clarifications and distinctions: learning from joint exploration as well as by the "rigors of research"; terminology of business and that of science; testing theory and testing materials for classroom use; normative models and clinical data. The minutes of this meeting continue:

> A worthwhile research project might be undertaken, Dr. Chamberlain pointed out, by inviting as consultants some behavioral scientists who have concerned themselves most notably with this area. They might meet with Southern Case Writers in a seminar period of approximately a week, after which participants could undertake the research and writing of a case to achieve the desired objective. It was his thinking that, at the end of this project, participants could meet again for about three days to discuss their efforts with each other and with the specialist who had served them originally.

A proposal along the lines discussed in this meeting was subsequently prepared by the Southern Case Writers Association, for which The Ford Foundation made a generous grant covering a two-year period beginning August 11, 1959.

Thus, a group who wanted modest help in getting out to write up policy-making situations in their region were tossed the hot potato of the decade — integration of science and administration. The Southern Case Writers rose to the challenge. Looking back on the proposal three years later, Dean Clive Dunham of Mississippi observed, "You know it really was a major innovation for a group of professors just to ask for consultants."

A Meeting of Minds

Following the suggestion of Dr. Chamberlain of The Ford Foundation, the Southern Case Writers Association secured three social scientists as consultants: Raymond Bauer, Ford Foundation Visiting Professor of Social Relations, Harvard University; Nicholas Demerath, Director of Social Research Institute, Washington University; and Howard Baumgartel, Chairman of the Human Relations Department, University of Kansas. The first meeting of the Southern Case Writers Association with the consultants took place in Tulsa, Oklahoma, from the 5th through the 9th of September, 1959.[2]

Because many universities are facing the problems of bringing professors of business administration and professors of social science together, this meeting has significant implications. The agenda allowed time for discussion around several presentations. In addition, at least one oral or written case was brought into the group for discussion each day.

(1) Dean M. M. Hargrove, Chairman of Southern Case Writers Association, "Case Writers' Need for Organizational Theory and Knowledge of the Behavioral Sciences"

(2) Dr. Raymond A. Bauer, Ford Foundation Visiting Professor, Harvard University, "An Approach to the Concept of Rationality in Individual Behavior in Organizations"
________________, "A Generalized View of Control of Organizations"

(3) Dr. Nicholas J. Demerath, Director of Social Research Institute, Washington University, St. Louis, "The Firm as a System, with Special Relevance to Executive Decision Making"

(4) Dr. Howard Baumgartel, Chairman of the Human Relations Department, University of Kansas, "Problems of Scientific Method in the Study of Organization and Management (and/or) Leadership, Supervision, Group Factors in Organizational Effectiveness"

(5) Discussion Led by a Case Writer, "How Can We Use What We have Discussed?"

The subsequent results of this cooperation justify a frank report of the awkward first efforts. Even within the group of professors from schools of business administration, there were many differences which had to be aired before any consensus could take shape. There was the question, for example, of whether specialists in marketing, production, and finance could

[2] This was in the fifth year after the first Visiting Professors Case Method Program, from which participants moved to organize for further case work in the south. By the time of this meeting, 31 participants from southern schools, including 5 deans, had gone through the Program and 11 schools had sent 2 or more participants. The time to organize a working cooperative group merits attention because a somewhat similar group from midwestern schools of business administration has been working since the summer of 1963 to get under way.

teach a general management policy course. There were questions of whether behavioral science was something of concern to functional specialists other than those in personnel management, whether behavioral sciences represented an alternative to functional specializations, or what. Other professors of management held that policy making was only one function and had to be considered along with organization, control, and so forth.

The social scientists documented their presentations with substantial papers and reading lists. Regardless of the clarity of presentation, however, it was far from operationally clear just how a person would use the concept of "rationality," for instance, in a particular field work situation. The social scientists disavowed their expertise in business administration. The professors of business administration likewise questioned their own competence in utilizing the tools of the scientists. Both groups agreed in the course of these first few days that cases could help by providing data to be studied by both groups. One of the scientists acknowledged that the case writer might get more help from novelists, who had to create the total reality, while the social scientist was seeking to keep everything fixed, except for one or two variables, so that he could experiment with them and explain their behavior. It was repeatedly brought out that the administrator is not dealing with a controlled situation.

Attention then turned to how a problem is recognized. Here the sociologists' concept of the firm as a total system had appealed to the group in the abstract, but the need for a way of thinking about side effects of action became apparent. Professor Baumgartel's cataloguing of concepts was not readily acceptable as an effective categorization of administrative issues. The scientists' desire to have the professors of business administration respond with some categories of policy cases took the group back into the ambiguities of marketing, production, finance, and other subjects, as categories around which to integrate the contributions of the social scientists. This fragmentary account may be enough to indicate the difficulty of putting together a program of joint activity.

There was some general agreement that the behavioral scientists promised to be of maximum use in determining means when the ends had been given. It also was agreed that a major job in the business policy course was to prepare students as business executives to identify ends. At this point, the behavioral scientists and the business school professors saw that the needs of the enterprise as a whole and the purposes of individual executives were different. These differences had to be identified from case to case. A start was made in the second conference in mid-July 1960. Presentations by the three consultants were reviewed in order to clarify their points of view further. Then there were new case presentations and individual appointments with consultants.

During the next few years the case writers met with the social scientists in the spring to plan the case research efforts and then met again in the summer or late fall to discuss what had been written. Typically these cases were duplicated and sent out to participants in advance. The participants were divided into three sub-groups, each meeting in turn with one of the social science consultants, to discuss the new cases. Learning to evaluate and to utilize the insights from new viewpoints was a major value of these meetings.

A Working Relationship

In the six years following the meeting at Tulsa, southern professors worked diligently in the field studying policy issues with the encouragement and professional criticism of social science consultants. Some of these professors in schools of business administration were themselves well trained in the social sciences. One member of the editorial committee, for instance, had been in charge of directing social science research in one of the military services during the war. Several of the professors held Ph.D.s in economic theory. Likewise, the social scientists who had been retained with the help of The Ford Foundation included men who had considerable personal experience in administration. Nevertheless, it took three or four years of working together before a productive relationship emerged, an important indication of the state of the arts of administration and social science. The eventual pattern was one of alternately considering case research plans with social science consultants in group conference, doing field work and case writing individually, and then discussing rough drafts with consultants and small groups of case researchers. As the cases were completed and contributed to the Intercollegiate Case Clearing House, a special effort was made to get each author to add to the abstract of the case at least a few words on how its use might reflect the social science interactions. Eventually teaching notes were prepared for instructors using the printed case book.

Although attention has been directed to the efforts of the Southern Case Writers Association to bring into focus considerations of organizational theory, social science, business functions, and business policy through case development, it is important to note that the administrative innovation of the Association was no mean feat. It is an unusual accomplishment to produce anything by 32 professors — even professors of business administration. Because many other schools are concerned about the costs and organization for case research, the experience of the Association with the mechanics of cooperation may be of interest.

Chairman Hargrove noted with his characteristic modesty and frankness after two years of operation under The Ford Foundation grant, "Through the editing activity, there had been formulated a system of *guidelines* which would have aided our activities if they had been formed some two years earlier." For really pioneering effort, guidelines may not be as useful as discerning hindsight.

Procedures had been established by the Executive Committee of the Association in their meeting during the orientation sessions at Tulsa in 1959. Grantees were not to be limited to alumni of the Program but were to include those who had evidenced interest in the use of cases.

(1) The Secretary is to draw up an authorization form to include: an abstract of the proposed case, the name of the company serving as the source of the case information, place for signature of the Applicant's Dean.

(2) Payment is to be made to the school upon release of the case proposal by the screening committee and upon receipt of a formal application approved by the Applicant's Dean.

(3) The amount of the grant is to be determined by the formula, 3/9 x base salary of the regular nine months' period.

(4) Lump sum check is to be mailed to grantee's University, to be paid out according to University policy.

(5) The individual University is to provide travel funds; if this is not possible under existing University policies, the Dean must requisition and justify the requisition on this basis.

(6) Travel expenses.
   (a) Allowance shall not exceed first class plane fare.
   (b) When mileage is allowed, the rate shall be 7¢ per mile.
   (c) All requests for travel payment shall be made to the Secretary-Treasurer and requisitioned by the Applicant's Dean.

The Executive Committee was also asked to screen proposals in order to prevent inclusion of too many cases from the same industry or government setting, and to prevent overlaps in case types. Subsequently, an editorial committee was created to dovetail the final cases into a publishable unit. The members were Deans Hargrove (Tulsa), Harrison (TCU), and Swearingen (Oklahoma State).

The experience of the Association was so promising that even while the first volume of cases was being edited, deans of the schools and The Ford Foundation agreed to a three-year program in which funds from The Ford Foundation (received November 22, 1961) would allow disbursements which would be matched by participating institutions. This endorsement by the deans as well as The Ford Foundation testifies to the care taken by the Southern Case Writers Association to involve the administration of each university in approving the participation of faculty members in a manner that fitted into the unique requirements of each school.

Implications for Further Cooperation

There was general agreement among consultants and case writers that it took a year merely to learn to talk with each other. The professors of business learned how the scientists thought, and eventually all three of the consultants became more closely affiliated with business school faculties. Some of the distinctions which were being brought out in Tulsa got carried on into the book of cases produced from this collaborative effort in the form of a "grid" which cross-referenced the functions of the enterprise and the functions of management. In this first publication of cases, the influence of the scientists at the Tulsa meeting was also evidenced by an additional "grid," which cross-referenced the functions of individual and group behavior. Although different parts of these grids were of different value to different elements in the group, they apparently provided a structure which allowed a heterogeneous group to get on with a common task.

One tangible result of the collaboration between the professors of business administration and the social science consultants was the book of 32 cases, *Business Policy Cases: With Behavioral Science Implications.*[3] The multiple considerations, "the grids," involved in the preparation of this case book were summarized as follows in the announcement of publication:

> Each case covers at least two of the functional areas of business. These cases provide a framework through which the student can analyze business policy in terms of the planning, organizing, evaluating, and controlling functions of management. The cases may be further analyzed in terms of the core areas of the business curriculum: accounting, business law, economics, finance, personnel and production management, and marketing. In addition, each case may be used to examine the behavioral science implications found in business operations.

Thus the original purpose of substantiating or refuting organizational theories of various disciplines was modified as the two groups worked together. As Professor Raymond Bauer, the senior social science consultant, said in another connection:

> . . . many marketers who have sought salvation in the behavioral sciences have come away muttering that all is chaos, and that the only answer given is "it depends." They overlook the vital fact that we have been finding out *what* "it depends" on. . . . I suggest an intermediate course between not using social science findings

[3] *Business Policy Cases: With Behavioral Science Implications,* Merwin M. Hargrove, Ike H. Harrison, and Eugene L. Swearingen, Editors (Homewood, Illinois: Richard D. Irwin, Inc., 3d rev. ed., 1969, in press).

> at all and their full technical exploitation. . . . Above all [the practitioner] must be cured of the notion that social science findings will ever give him an invariant "right" way of proceeding. The fundamental finding of the social sciences is that the preferred course of action is specific to the situation.[4]

This is not to deny the importance of the social sciences and their subdivisions, but to differentiate the tasks of generalization for the scientist and for the administrator. As reported in an earlier chapter, Dean Lombard in speaking to Program participants stressed: "We do not learn theory to deduce an answer. The proper role of theory is as a guide to inquiry, to the meaning of events."

The need for clarifying the roles of science and administration persists throughout American schools of business administration. The former Program participants in the Southern Case Writers Association have gone through a unique and valuable process of working through this differentiation. The cases in the book provide materials for classroom discussion, as Dean Hargrove put it, "to lead the student to think and act like an executive." The introductory text material and especially the guide for teachers provide additional insight from this cooperative experiment. Such cooperative experiments and the professional participants who contribute to these experiments should be encouraged to develop in more detail the distinctions they have found useful between the functions of an enterprise, the functions of the manager, and the functions of the social scientist.

In one of the meetings of the social scientists and business school professors, the point of view of the scientists was summarized: "Basically behavior is explainable." Work with the social scientists had clarified the conditions of experimentation and what the determinants might be. Without getting into the metaphysical issues, however, the administrator is aware that he frequently comes to a point in time when his action is not determined, but depends upon his discretion.

Paradoxically, as the case writers and social scientists worked

---

[4] *HBS Executive Letter,* February 1966 (© President and Fellows of Harvard College).

together it seemed that the more data a social scientist had about the situation, the less use he was to the administrator. For instance, the economist is not much direct help to the administrator in a competitive situation where there are many data available on price movements in a competitive market. He wisely tells the administrator to find out what the consumer basically needs and to use some distinctly different ways of meeting these needs. The administrator in that situation needs help in psychology and design more than in pricing. However, when there is little information about the market for a new product, the pricing economist may be of greater use to the administrator by helping to identify for him some of the limits within which discretion can be exercised in pricing, and to suggest areas of inquiry to foresee consequences.

This paradox suggests, among other things, that if the social scientist is to help the administrator, he needs to turn from expounding his knowledge of behavior of one or two variables under controlled conditions to developing ways to explore conditions in the concrete situation of the administrator in the time available before a decision has to be made. The relationships about which the case writer has most data for the scientist may not be of critical importance for the immediate decision facing the administrator, while the conceptual model of the scientist in the absence of data may give the administrator and case researcher strategic clues to explore. Clearly the experience of the Southern Case Writers shows that the social scientists have no panacea which will replace the administrator. Also, the experience shows that the social scientist can help by illuminating that point in the evolution of an enterprise when the professor of business administration, the social scientist, and the practitioner look together at the need for action.

The view of organization, from outside, by the scientist needed for his abstraction, isolation, and experimentation is related to but different from the concepts of organization needed inside the organization by the practitioner and the professor of administration searching for points of discretion and action. However, these outside-inside views need to be as compatible as the two sides of a coin. The Southern Case Writers Association experience gives some indication of the time and effort required for a group to understand that these differences in orientation exist and how they may supplement each other.

Few schools, if any, can have access to enough social scientists to get professional help on the specific knowledge or technique relevant to every specific case. Few social scientists will know where and how to find the professor of business administration to establish the inside counterpart to his outside concept of organization. Clearly much more can be done within many universities to exploit the Southern Case Writers-social science consultant prototype of a working relationship.

Chapter IX

Summary and Implications—Progressive Approximations

This book and the Program it describes were motivated by the conviction that our modern society desperately needs administration and that the development of orientation, knowledge, skill, and maturity for administration can be accelerated by the study of cases. The Program was not intended to produce research findings, nor to test a theoretical proposition. This chapter, therefore, is not a statement of findings but a summary of the operating model used in the Program and some of its implications. It is addressed to those who are concerned about the estrangement of active administrators, academic specialists, and the coming generation (students) in our society. It does not attempt to prove anything, but to suggest some promising directions for further exploration. It does not point to a panacea, but to a basic common reference: for the administrator in his enterprise; the professor in his subject; and the student in his revolt.

The Program was offered to allow professors and deans from a cross-section of universities to experience for themselves a set of activities and relationships made possible by the study of administration in business by cases. Student revolt has become widespread enough to dramatize one element in this set of activities and relationships: the student revolts against the irrelevance of the professor and the unresponsiveness of the administrator, the "establishment." The anxiety of the "establishment" is widespread enough to identify the importance of the administrator as another element in this set of relationships: the administrator feels under attack, and misunderstood. The isolation of the specialist, represented by the professor as the third element in the set, is voiced repeatedly; the professor feels isolated and ignored.

It is no mean feat to find something in common for these three diverging forces in our society. The unifying reference is not any one complex or sophisticated machine. The unifying idea is deceptively simple: Look at the data requiring discretionary action in a concrete situation.

In emphasizing the interrelationships among the administrator, the professor, and the student, the "worker" is not being disregarded. A different axis is emphasized — different from the distinction between management and the worker. Attention is being directed to processes in which all men are involved, whatever their titles. For convenience the terms administrator, professor, and student are used. For some purposes these terms also stand reasonably well for the processes called administration, specialization, and learning. However, it would be grossly misleading to think that only persons formally in those roles are concerned with these pervasive processes.

Much of the impact of the Program upon its participants came when the use of cases actively involved a man in all of these roles in all of these processes in a way which allowed him to compare his personal insights directly. Seeing the same case as part of these three processes from these three points of view brought together feelings and beliefs previously ignored or kept separate. For instance, there is a difference between the professional or occupational orientation of the administrator toward his enterprise and the professor toward his subject, and the contrast for both between the orientation he feels toward his job and the identity he feels as a person. This distinction exists regardless of job classification, whether it be Board Chairman or the least skilled laborer.

The novelty of the Program as an operating model was not the newness of the ideas, but the use of them in this set of relationships, which in modern life are all too often isolated. The search for data where purpose and decision are involved in actual situations, and the use of these cases with students, involves a person in all three processes: administration; specialization-isolation; and personal learning, growth, and maturity.

SUMMARY

In summary, the case method has been effectively used in widely different institutions, depending upon the extent to which several important conditions prevail: (1) the availability of kindred spirits in the academic community for criticism and support; (2) the provision by university administration of time, money, and personal encouragement; (3) the access to enterprises and management for continued case research; and (4) the interchange of case material.

The basic condition for extension of the case method is commitment to the assumption which governed the Program: the critical importance of administration in modern society. Hence the significance of studying it. The case method has been described as more than a device in the classroom. It has been seen as personal experience in an interdependent set of activities and relationships bringing together active administrators, professors, and students in a common process of learning. Participants in the Program concluded that the unexpected consequence of this involvement was seeing themselves more clearly. Insight came from work with administrators in discovering case data; from writing the case with data separated from the participants' own opinions; from sequencing cases and concepts; from response of students as they projected themselves into administrative situations; and from reconsideration of criteria for selecting, evaluating, and criticizing students and academic colleagues.

As has already been noted, one of the major conditions of the Program was the significance of personal involvement by the professor in many relationships. The willingness of the individual to continue this process of stretching in many directions is therefore a major factor in the extension of the case method. Discovery of the case as a point of intersection, of interchange among these processes of specialization, administration, and learning gave new enthusiasm to professors participating in the Program.

The coaches were also a unique feature of the Program, but their function has been provided in a great variety of ways under other circumstances. For instance, the cooperative work of the Southern Case Writers Association with consultants has been reported in Chapter VIII. Also, the use of research assistants and graduate assistants was described as a division of labor to help provide critical detachment as well as active involvement in a situation required to get primary data.

The Models: Operating and Symbolic

From early trials and errors in the 11 Programs and from the experience of predecessors, participants could be brought into actual operations of the case complex with a very minimum of explanation. Results year after year confirmed the observation that in learning to use the case method, "the way to begin is to begin." Scores of professors learned by working in the operating model with increasing enthusiasm and skill. To repeat, experience in using this operating model rather than analysis of a theoretical model brought about this result. At this point, however, elements of a symbolic model may appropriately be presented as a partial summary, but the purpose of this final chapter is not to go through the potential interrelationships.

Among the interdependent elements in the study of administration by cases would be at least the following:

The Study of Administration

Roles:

Administrator	$= R_a$
Professor	$= R_p$
Student	$= R_s$

Processes:

Administration	$= P_A$
Isolation-Specialization	$= P_I$
Teaching-Learning	$= P_L$

Operations:

Observation-Interviewing in the Field	$= O_f$
Writing data for cases and opinions for journals	$= O_w$
Design of course sequencing of data and concepts	$= O_d$
Classroom discussion of cases and concepts with students	$= O_c$
Selection and evaluation of students and professors	$= O_e$

Values:

Occupational orientation	$= V_o$
Personal identity-maturity	$= V_m$
Knowledge	$= V_n$
Skill	$= V_l$

Case:

Issue, data, time	$= K$

Unknown:

You and I as whole person	$= X$

Identifying these variables may serve several purposes: (1) to show the complexity of even the simplest administrative situation; (2) to note that many a scholar has spent a lifetime and more lives will be spent trying to "understand" each one of the elements with which the administrator deals; (3) to note further that all these elements are interdependent, probably in exponential functions with a time delay of one or two generations of genes and generalizations; (4) to suggest to the mathematically inclined why this report has been written in terms of successive approximations to congruent involvement in the total set of activities and relationships; and (5) to remind all readers that in administration *purpose cannot be taken as given* and that the critical unknowns are human beings in the process of becoming.

This book, like the Program, is organized around operations.

The multiple values, emerging from comparisons of ones own reactions and those of other professors, to data of the case under different relationships, generated a tremendous enthusiasm. Such enthusiasm continued as participants extended their activities in a variety of institutional environments. It tended to decline when the richness of the matrix was thinned down by what could be called overemphasis on $O_e = R_p.P_I.V_n.$ X. The price of such oversimplification may well be student revolt and isolation from the vitality of the active administrator. At least 200 other equations are needed in the complete matrix. Although contributions useful to the administrator will continue to come from those who arbitrarily isolate some piece of life for mastery, as does the biologist, engineer, doctor, or any other specialist, contributions also will be made to the understanding and development of administration by those who get involved directly in the confusing but creative process of discovering cases.

With a final plea that no one delay long in getting out to where the action is, the reader is urged to start from wherever he is to see for himself what happens when he discovers new direction emerging from the interaction of data and discretionary action.

The Development of the Administrative Point of View

The basic premise of the Program and of this book is the importance of administration and the effective study of it. Need for administration arises when customs, rules, forms, or principles come into conflict. Individuals and organizations then face choices and must create new courses of action to maintain their integrity, to grow, and to achieve their purposes. Alertness to such need for action is a first step in the study of administration — for the man on the job, for the professor-researcher, and for the student. For all, the next step in the learning process is marshaling the evidence in a way which will get participants personally involved in taking responsibility for analysis and action. The power of the case method of learning stems from its relevance alike to the practicing administrator, to the professor, and to the student. The parallelism may be obscured, however, by different words when executives discuss exceptions, researchers develop a subject, and when students want to be in on the action.

It is important to study administration as a process, but it does not take place in a vacuum. The context also must be studied — where administration takes place — in business, in hospitals, in government agencies. One of the unfortunate results of the isolation that has so often taken place in academic analysis is that "business" is thought of only as the economist's abstraction of profit measurement. Attention to the total administrative situation and all that goes into the determination of policy direction could help those who are scornful of an administrative career to see it in a different light. Almost everyone in the United States is involved in the business system. Even in the most primitive developing country, the objective is to have more and more people become involved in the business system. We are not concerned about the terminology, whether it be called the industrial system, public sector, private sector, or people's project. It is the way in which members of the society share their labor and talent to meet human needs. In the United States, business enterprises employ all types of men from moron to genius — from grammar school dropouts to Ph.D.s. We need a way of thinking about the process of administration which can be shared by all these citizens of all ranges of interest and levels of ability. Particularly needed is a way of thinking about how changes to incorporate the gains of science can be brought about and how new activities can be made routine. In order for this process to be made comprehensible, attention must be given to the administrative components of every job.

Although the administrator is frequently characterized as a person who works through others, the person and the process must be differentiated. Surely this element of administration is crucial and an essential requirement to thinking and feeling about the organization through which one must receive communications and extend them in order to make and execute decisions. It also is important to recognize that everyone in an organization has some measure of administrative responsibility. There is an administrative aspect in *every job.* Therefore, we talk about the "administrative point of view." The administrative point of view is manifest in alertness to exceptions and in taking responsibility to make them opportunities for growth, for reinforcement of direction, of routines, of purpose. As purpose is modified, there is search for improvement in routines, customs, and systems. Awareness of a worker's contribution to the life of an enterprise and his benefits from the enterprise helps to give meaning to routines and changes in routines. At higher levels of administration, the meaning of transactions to the persons directly involved tends to be lost in abstract totals. But numbers are only transactions and people, out of sight.

It is useful, therefore, to distinguish between the administrator as a person and administration as a point of view. Persons with administrative titles and responsibilities do not necessarily have the administrative point of view. This point of view is an orientation which a person may have regardless of formal position. In most enterprises in the United States, the potential for it is the basis for promotion, and therefore is an important part of training for professional administration. In cultures where promotion to administrative position is at present more largely influenced by family or other considerations than in the United States, there is even more need to distinguish the point of view from the person or particular positions.

Meanwhile, the lack of congruence between the educational system and the institutions of society in which men must work transfers to business enterprises themselves a massive educational task. Over and beyond the inevitable job training, there is an ever-expanding concern with supervisory training, foreman training, sensitivity training, management development, and exposure to the arts and the perspectives of great ideas.

Clearly, the function of the educational system is not solely to train young people for jobs. Rather, its function is to develop the capacity of each individual as a person. As young men and women move from the educational system into the business system, the crisis is precisely the lack of disciplined intelligence, controlled energy, and emotional maturity, rather than ignorance about the job. Development of these personal qualities is a broader educational objective than knowledge alone.

Emphasis on the administrative point of view also makes the

reason for education more immediate. Emphasis on career aspects by contrast leads to the assumption that education is for the future — 15 or 20 years hence when the student becomes an administrator. Experimentation with a course or curriculum can be exciting on the assumption that administration is a process in which everyone is involved to some extent — now. The administrative point of view, as a way of looking at a situation for everyone — now — can get students into a frame of mind for taking responsibility. Men who will take responsibility are in such demand that they might well be made administrators within months instead of years. Happily, the point of view which makes learning a present experience also promises to make the process continuous.

There is an assumption in the administrative point of view that the individual can make a difference. In some cultures, determinism and reliance on supernatural rectification of man's lot are stubborn factors in the situation. Determinants and the place of discretion are not fully clarified in American thought on the art and science of management. In this connection, the mere fact that the Intercollegiate Case Clearing House in the past decade has received 7,000 reports of situations in which someone felt he had discretion to decide on a course of action is significant testimony to a widespread sense of responsibility and existence of choice. Credit goes to those pioneers who are getting cases from actual administrative situations and using them in their courses. The need to identify administration in business as a legitimate field of study, and to determine an appropriate way to study it, is great. Participants in the Program were impressed as much with the implications of accepting the study of business administration as an objective as with the refinements seen in the case method itself. Stress is still inherent in administration, because it deals with each situation as a living whole. It comes from the continual need to balance multiple noncommensurate values. There is great temptation to oversimplify the situation by taking only one value and maximizing it. Understandably, such stress carries over into the teaching of business administration, which must be concerned not only with all the multiple noncommensurate values in the administrative situation, but also with a new generation of students at a critical period in their growth toward personal identity.

The case method emerged out of the trials and errors of studying business administration with all these considerations in mind. It evolved from a determination to keep together the arts of judgment and pursuit of knowledge; to recognize the interactions of ends and means, of routines and exceptions.

Points in common can be seen in the administration of all kinds of enterprises: hospitals, school systems, the military establishment, the post office. Increasingly, the business enterprise and its manager appear as phenomena for study by the sociologist, psychologist, anthropologist, political scientist. All these areas of specialization are relevant for business administration. The task is to find boundaries or some subdivisions within which men may still learn to take responsibility for the enterprise as a whole.

The reconciliation of responsible action with the careful study of scientific laws and systematic relationships is part of the task of interpretation. It is difficult to avoid take-off into more and more abstract levels of discourse when there is no felt need to make the study relevant to actual administration.

The Case as a Means of Learning

The case truly is not just a device for instruction but a means of learning for administrators and for professors, as well as for students. The case is also evidence that the professor has found a new use of the ideas in his course to expose the critical conditions in concrete situations, a new balance with exposition of abstract generalizations. In turn, students find that use of the course concepts help them to discover more meaning in the data of the case when they try to take relevant action on the issue. Discovery of one point at which to exercise discretion will not solve all the problems of the world, but will help students tie into the creative forces of the world; to become factors in its evolution.

The practicing administrator quickly grasps the idea of a case, because it is roughly analogous to an exception. The professor, like other specialists, more or less explicitly thinks in terms of a system. He admits that exceptions may occur, but they are to be handled administratively. The idea that administration of a system might well be studied by a sequence of such exceptions — cases — therefore really is not so novel. To use cases does mean that the specialist can not go so far out toward the logical limits of his subject because he becomes more involved where the system of his special interest interacts with others, with the administrative effort to determine where the organization is going. By building his course around such cases the specialist becomes involved in determining objectives which he otherwise takes as given. The professor thus often finds new enthusiasm in the creative possibilities of his systematic analysis. The administrator who finds a specialist who is able to use his specialized system of concepts to explore emerging conditions in actual cases gains new power to clarify assumptions, preferences, and tentative purposes by predicting probable consequences.

Every administrator is working on many projects, items, crises, and deals for which data are being accumulated. Many academic researchers are also gathering empirical data. The case as discussed in this book typically combines the relevant data selected and summarized to permit rational analysis and a more or less explicitly stated issue requiring projection of a decision by students within three to six hours of study and discussion. Issue, data, and time are critical elements. What may make a decision difficult for the administrator may be so personal that the case does not challenge students, and what appears to be a simple choice for the administrator may evoke such strong personal feelings among students that the case is difficult to handle. In both instances, the collection and presentation of the data may have met the highest objective standards of quality. Although the case can provide a common reference, misconceptions of the case, especially the issue, may obstruct learning by students, course design by professors, and determination of policy by administrators. Not only relevance, but relativity of time and perception is involved.

This relativity of perception (or reader interpretation) is a shock to the new instructor who frequently is not prepared for the time required for the class to sort out what is personal to the reader, what is peculiar to the point of view of the administrator or of the specialist, what is subject to further analysis by more precise knowledge of the course concepts, what can be clarified by more skill, what a person can accept only by developing personal maturity, and what can be done now.

Design of a case-course is a continuous process of searching and sequencing to discover the mix of cases and concepts which meet the needs of a particular class of students. As cases are presented where alternatives have not been formulated, or issues have not been clarified, it is more difficult for students to see what is involved. Most difficult of all is a situation where only a general uneasiness is felt and must be diagnosed by careful and expert analysis.

By building a course around cases which bring into question the pre-science assumptions and what the specialist usually takes as given, the professor often finds new enthusiasm in the creative possibilities of his subject.

In the discovery and selection of data in a complex situation most relevant to the study of administration, it has been useful to think of the case as a joint product. It is not something ready-made in the desk of the administrator to be had for the asking. Nor is it something created at the desk of the professor. It is not just any empirical data counted in a survey. The case emerges from the professor's efforts to understand the administrator's experience with things, people, and events, in which the ideas of his course might be relevant, particularly at the point where the facts call for discretionary choice, the point at which the issue becomes evident.

An issue may be implicit in a situation long before it becomes evident, or an issue may be foreseen long before data are available. Much of the skill of the experienced administrator is in bringing the resources of the organization to bear on a creative resolution of an issue. Foresight enables differing points of view to be expressed and taken into account in face-saving accommodations. Months and even years may be taken in working out courses of action which preserve the enthusiasm and contributions of diverse groups. In a group of specialists who must continue to work together, the polarized confrontation which proves so thought-provoking in the classroom is not necessarily the model for handling an issue. However, the issue and a sequencing of issues helps to establish the administrative boundaries to a field of inquiry.[1] The need for decisive action implicit in a concrete set of facts provides a powerful reference point. Anyone can go off into any of the infinite subdivisions of abstract knowledge or practical experience and always have a meeting point to which to return — the point at which his thinking can make a difference to the action needed in the situation at the time.

The sharply defined issue which so effectively involves a large class of students, however, is not necessarily what the professor, researcher, or alumnus should expect to find in a healthy enterprise. In fact, the experienced administrator seeks to identify as early as possible potential needs for change in policy direction in order to avoid the emotion-laden "policy issue" discussion which has proved so effective in class. It is as though his experience with the divisiveness of policy issues enables him to foresee the need for change. Hence he is able to enlist differing points of view and to incorporate them into a course of action which meets the manifest needs of the situation when the time of action appears.

Thus, it may be unlikely that the case researcher or recent graduate sees in a successful business any situation that looks like the case he saw so dramatically discussed in class. Here is where a person's knowledge of his subject helps to identify turning points which management has worked by skillfully in the past, or to foresee (with management) turning points appearing in the future as a result of exceptions now appearing or of changes in the environment about to take place.

It is the time aspect of living concrete situations which most often is distorted to bring the experience into a class period for involvement of students. Similarly, it is adjustment of time

[1] ". . . when man tries to understand he generally loses sight of the problem he meant to study," p. 31, *Human Destiny*, Lecomte du Nouy (New York: New American Library of World Literature, Inc., 1947).

which is so important for students going on into enterprises as potential administrators. It is imperative that the distortion be rectified in transferring issue analysis to alertness, foresight, and planning.

The need to capture the attention of students and to get them fruitfully involved in an administrative situation within a block of three or four hours helps to shift the format of administrative cases away from the precedent style of legal cases or the clinical diagnosis-and-treatment style of medical cases. Since administration is characterized by the unknowable future, the case format turns from where the action *was*, to making students take personal responsibility for analysis, purpose, and action *now*.

There is danger that by disposing of issues in hours, students may not hear the professor's caution that establishing the routines of the new policy may require months and years of patient accommodation to the needs of the organization; in other words, the identification and reinforcement of precedents and effective custom. Not everything can be learned as well in the classroom as on the job. But even on the job, training now sees economy in attention to exceptions as well as repetition.

Work with students on cases provides access to the all-or-nothing over-simplification which so often leads students to obsessive thinking. Cases involving the conflicting occupational orientations of the specialist and the administrator allow the professor access to the struggle of the individual for a personal identity apart from the momentary demands of a particular job. Few if any jobs are all administration or all specialization. Even the most routine clerical tasks involve an immense amount of discretion, as was discovered when trying to program them for a computer which takes only a yes or no answer. Even the highest administrative positions may be looked upon as personal stepping stones, and the incumbent is only in part speaking for the organization, in part articulating personal values.

Use of cases in the classroom provides opportunity for instructors to treat students as mature, responsible citizens concerned with taking a constructive part in society and, at the same time, recognizing that they are becoming a part of that society. It is a way of avoiding the sad dilemma of the college student who wrote, "I guess there's little room anywhere for *growing*, rather you are just expected to be *grown*." [Author's italics.] Out of the struggle with hundreds of actual cases, students gain confidence in their power to discover constructive action possibilities. They discover that they do not have to carry the weight of the world, just work with it.

The Program gave professors an opportunity to see for themselves the complexities of concrete administrative situations, to fit such experiences into their own course concepts, and to multiply this process for large numbers of students by preparing a case in written form and using it in class. This set of activities was described in earlier chapters in this book. Insights of the participants were clarified by contrasting reactions to the same case data in these different activities, processes, and relationships. Too few see the common interest of the practicing administrator, the potential administrator, and the professor of administration in business or some function of it. Although this book describes, from a professor's point of view, a whole set of activities, these activities also involve practicing administrators and students who may be potential administrators.

To some extent, variations in emphasis will come because no one man can do the whole job of educating a group of students to take responsibility for administration in business. Variations exist also because some assume that time is saved by giving students information and ideas in a logical sequence before they are given opportunity to use them. Others assume that random access programming through cases using varied concepts might well accelerate the student's capacity as an adult to use information and ideas and to continue to enlarge and renew his resources.

Faculty members offer students different models in using skills of analysis in case discussion to distinguish between a system of classification, a predictive proposition, a normative statement, and a general model or theory of how a relationship works. The course made up of cases is an invention which makes possible the further exploration of the territory with the existing state of theoretical development. It also allows for cultivating the disposition to decide, but does seem to thwart the desire of some faculty members for elaboration of the conceptual framework. There is a never-ending search for ways of talking about theories of administration and developing them without losing the disposition to decide — without shifting from the administrative point of view to one of abstract theorizing.

This search for balance becomes an issue for a professional school in a way which is not felt in a Graduate School of Arts and Sciences. There the graduate students are in process of becoming professors themselves. MBA students are in the process of becoming practitioners. A professor may elect not to involve himself in some of the multiple commitments which he faces, but that in itself is a position on the point. Noninvolvement may mean that he becomes progressively less relevant for the student of business administration.

Implications

The reaction of participants in the Program suggests that one of the dynamics of the case method is the urge to find congruence in personal responses to case data by the administrator in the situation, by students in the classroom, and by colleagues in professional discussion. These gaps between generalizations of the subject and observations of concrete situations drive the researcher on to further investigation of the facts and reconsideration of concepts. Many of the participants had had very little firsthand experience with administrators in business before coming to the Program. Exposure to cases gave them an idea of where to look and what to look for. Participants who had had personal experience as administrators in business found in the exposure to cases an idea of how to present data of experience in a way to involve others actively and to avoid repetitious accounts of "my experience." Thus the use of cases helped to make a professor's course an instrument of learning for himself as well as for his students. The use of cases gave professors access to the intangible, subjective, elements of learning and personal growth which have the longest periods of gestation — and which all too often the professor ignores or leaves to future experience, while concentrating on exposition of his subject. Use of cases helped the professor involve his students in purpose identification as well as problem solving.

In some ways the use of cases complicated the professor's coverage of his subject. But in other ways the use of cases permitted the professor to convert a subject from static pigeon holes into which pieces of life are put, to an organic system of concepts which can be used to see life more clearly and as a whole. As a specialist the professor rightly holds that others are responsible for the direction of organizations and for their own personal lives. By searching for the experience of administrators with his subject, however, the professor (specialist) may discover where the determinants in his subject force choices or respond to them. By redesigning his course around such issues, he helps his students assimilate the subject in terms of fixed and discretionary elements, and to grow in maturity.

The case method is a way of keeping these points together so that what goes on in the classroom may be subject to correction by observation in the next field case situation. Progressively as the professor goes on in adjusting his position there is a continuing approximation to a balance which is functional for his learning and that of his students.

The very interests which drew a person to the case method program provided both basic motivation and pervasive uneasiness. It is a challenging responsibility to study administration in a way to involve the student as a whole person in a process of growing:

From immaturity to identity as a person and as a professional administrator
From ineptness to skill — in discovery — in relationships
From ignorance to knowledge — knowledge of policy formulation in an enterprise, and knowledge of subject and discipline.

Most professors acknowledge responsibility for helping students move from ignorance toward knowledge. There is less agreement as to the role of the professor in the other phases of the potential administrator's development toward wisdom. In education as in other processes of growth, there may be pathology as well as normal process. For men from scholarly backgrounds the pathology of impulse, "from the seat of the pants" response, is quickly apparent. For those exposed to large organizations, the pathology of indecision is impressive. As in sailing, the controlled opposition of sail, keel, and rudder leads to progress; in the case method, the controlled interaction between knowledge, skill, and identity makes for dynamic growth. The introduction of cases into the pattern of a professor's relationship with his subject, with his students, and with a particular enterprise leads to progress in a state of moving equilibrium. Since these systems converge only at a moment in time, the manager of a business firm needs access to many scholars. Conversely, scholars need access to many managers. Professors benefit by access to many students, and vice versa.

The rationale for the Program was that progress comes from successive approximations. The objective, therefore, was not perfection of any one activity in the case method pattern as much as getting men involved in the whole set. The director and coaches of the Program avoided opinions on whether a participant had written a "good" case or had conducted a "good" class discussion. Instead, they cajoled the men into going on with the schedule, to deepen their involvement in other activities of the set around a body of case data.

The schedule also required participants to make choices in various situations. Thus, men were re-examining their values in specific terms and naturally were extending the implications to more general import. To avoid getting hung up at the obsessive unanswerable level, the Program tried to stimulate reconsideration of men's values by feedback from the group around experiences in the field, in case writing, course design, classroom discussion, and student evaluation. The Program also tried to avoid over-thinking imponderables by getting men to take their case data on to further involvement in other stages of the set.

The case data thus provided material for multiple reference, a bench mark, against which a man could clarify his orientation and knowledge as a specialist, or his orientation and experience as an active administrator. Discovery of the issue for action and the meaning of the data showed the need for logic and also the need for a different mode: inspiration, situational thinking, skill. The process of discovery engaged the researcher as well as the administrator, the students, and the researcher's fellow professors in unexpectedly different beliefs and often aroused surprisingly strong feelings — unacademic defensiveness, hostility, or fear.

Every time a professor went into the field he obtained insight into determinants influencing the administrator. He eventually found also the point at which in the whole system of determinants there was an adaptive moment when discretion by the administrator could help to create a new direction for the enterprise. When the professor brought back the firsthand data from the field and began to put it into the written case, there was opportunity to examine the way in which his own opinions had become involved with the data so that in writing the case he was also examining his own concepts, values, and beliefs. This examination was further extended when the professor began to fit the written case into his course. It sharpened his use of concepts for exposing conditions in new situations as well as for stating the general relationships under given conditions. Then each time the case was used in class, the reactions of students brought further insight into the process of learning by which the individual grows in skill, maturity, and identity as well as in knowledge. Thus, the involvement of the professor in all the elements of the case complex led to successive approximations, toward an ever-moving balance, his own case method pattern.

Participants almost uniformly expressed appreciation for having been "scheduled" through the Program in such a way as to feel some exposure to such range and depth of experience. A few were frank to admit that this sampling was enough and that they were going to concentrate on an aspect of the administrative process in which they felt comfortable. Others were willing to try another cycle or two in their own university contexts.

Implications and Insight

It was the experience of a *total* set, cycle, or pattern of relationships which participants valued. In 1962 The Ford Foundation made a sample survey of 58 participants and deans to help in deciding whether or not to continue support of the Program for 1963–1965. The Ford Foundation officers were impressed that the participants recalled the Program as an opportunity to take part in a complete cycle, including field work and case writing. It resulted not only in their increased use of the case method but also in their increased awareness of teaching methods in general and reappraisal of their own teaching. On various details of the Program and its length there were mixed reactions and numerous suggestions. Beyond the professional values, the comments on the Program's great personal value to participants were especially noted in The Ford Foundation survey.

This unintended personal consequence of the Program seemed to come from the accumulation of related activities which required active personal involvement. The feedback gave the participant not only professional criticism but also an unusual opportunity to see himself:

(1) In relation to a specific enterprise and the executive responsible for action in it. People were reacting to him as well as to their jobs and to the organization.
(2) In the search for action points in the situation where his subject knowledge might make a difference by exposing specific conditions making generalization about his subject relevant to policy direction in the specific enterprise.
(3) In sorting out these complex elements into a statement of data requiring action, "the case;" and a separate note of his own opinions and ideas for utilizing the case in a course sequence for classroom discussion.
(4) As an instrument in the process by which students vicariously go through the learning experience which he had in the concrete living enterprise.
(5) In relation to other faculty members, responding in different ways to the objectives of education for business responsibility and the criteria for career advancement in the academic community.

The case method of thinking about a man's special subject leads him to the points in a specific situation in which the subject makes a difference. It leads to restating abstract generalizations as search patterns to help find conditions which can be modified in the concrete situation. The academic world is not generally organized to appreciate this method of testing analytical concepts by action in concrete situations.

Using a case to illustrate a topic in a series of lectures or discussions of a subject is quite different from accepting the case as a map of an emerging situation in which a person in the situation can make a difference by his analysis of it. This difference in the view of the case means a reconsideration of a man's existing ideas about his subject. The case is the acid test to see if the concepts in the course provide distinctions which make it possible to do something useful under the conditions in this particular situation. It means a reconsideration of the instructor's ideas about what and how students learn. It means a reconsideration about ideas of the administrative process and personal values.

Rethinking views across this broad range of involvement is more basically disturbing than most men anticipated. They found that getting out into the field and identifying actual administrative situations, writing up the data for use in class, and then leading an analysis to a course of action made a person keenly conscious of many assumptions, values, and habits which had to be reassessed. It was this experience of self-examination and development which the participants reported as one of the major benefits of the Program in The Ford Foundation survey at the end of 1962.

Institutional Context

Men who had high career expectations in their schools were nominated and selected to participate in the Program. The fact that they came from so many different schools, and that at the end of 11 years, 77% of the participants, other than those who had died or retired were still in the same institution [all but 4% were still in education] indicates that in the United States there is a broad cross-section of institutions in which professors may make a career while studying business administration by the case method.

The institution influences the use of the case method in bringing together a particular group of students and a particular group of professors. The image of the institution, the recruiting practices, the success of the graduates, all of these are important elements in the effectiveness and the character of the case method pattern of any individual professor. The institution also conditions development of the case method by the way in which the curriculum is organized. Rarely does a group of students come together for a sustained, formal program in which only one professor is teaching. One must consider therefore how the parts of a total program for students are arranged. The effectiveness of any individual professor in his use of the case method is conditioned not only by the subdivision of the program content but also by the colleagues, his peers, who are drawn together in the program as a reference group.

The concern of Program alumni about transfer of experience to their home institutions can be summarized into three broad categories: (1) support for further case research; (2) integration into other classroom activities; and (3) impact on personal professional development.

> Many were anxious to compare experiences in getting, or not getting, support for further field work. For all it was a question of time. For many it was a question of funds for travel and research, editorial and stenographic assistance.
>
> The study of administration by cases was supported, supplemented, or supplanted by in-basket exercises, business games, simulation, programmed texts, computer analyses, to say nothing of departmentally authorized text books and the irresistible urge to lecture.
>
> Appreciation or antagonism of colleagues, especially when expressed by promotion, influenced allocation of effort.

Reactions of individuals typically cut across all three of these categories.

Correspondence, questions, and observations made on visits to different institutions suggest a gradation of transfer. At one end of the continuum were men who personally were very much involved in new field work. Enthusiasm of alumni who were able to continue actively with field case research and development has been striking, over the 15 years since the Program began. Some of these alumni returned in subsequent years to talk with current participants about their experiences. Some of their efforts had been well financed and substantial time made available. Others had had to rely upon the help of current students or recent graduates as case associates. Some had been able to enlist colleagues from other disciplines in a team project. Chapter VIII explained briefly how the Southern Case Writers Association had been organized after some years of informal meetings in which former participants of the Case Method Program had become acquainted with each other, and described subsequent progress in collaboration with social scientists as consultants under grants from The Ford Foundation. Later other participants formed the Midwest Case Research Association, members of which have been progressing on a modestly financed program. For those who were thus actively involved in field work with actual business administrators, there was a constant reconsideration and modification of the curriculum to accommodate relevant new cases.

Further along this continuum were men who were developing their own courses by selecting cases from the Intercollegiate Case Clearing House. Some of these men were putting a number of their own cases together for publication. Others were calling on the Clearing House for materials designed for courses for specific programs. These men were concerned with the divergent qualities of the material in the Clearing House and the frustrations of having access to a field situation only secondhand through the eyes of someone with a slightly different point of view.

Further along came a fairly large number of men who were using printed case book collections. They had problems of utilizing them in undergraduate programs and in case-course development for graduate programs. Some of these men also were concerned about the use of such case-courses in executive programs or overseas.

Then there was the group along the continuum who were concerned about the introduction of a few cases into traditional text book or lecture-discussion courses. These men were concerned about the mechanical problems of getting students to buy both cases and text books and about the development of combination text and case books.

More than two-thirds of all the 227 participants had come to the Program from schools where further case development depended upon individual effort, with very little more than a permissive attitude from the administration. However, the number where active support is available is increasing.

Rarely does one professor have both the time and the resources to develop all the material for his own course. Nor are his students fully ready for what he wants to teach. In the transfer of the case method to institutions with limited facilities for support, therefore, the experiences of alumni of the Program in a cross-section of schools illustrate the possibilities as well as the frustrations.

This report on the contagious enthusiasm of the alumni who in one way or another were able to extend actively their field work does not mean that they no longer had questions or uncertainties. New stresses with more experience in the whole complex of case activities brought new perspectives. The use of cases to study administration, as repeatedly pointed out, does not create such stresses but does provide constructive access to them.

Progress or Pathology — "Stability is But Balance"

Mr. Wallace B. Donham, the second dean of the Harvard Business School, used to quote from Robert Bridges' *Testament of Beauty*, "our stability is but balance." Approximated but never reached in the study of administration, or in administration itself, balance comes in the creative, never-ending search among divergent objectives.

In Appendix 4 are summarized a series of discussions about the case method held by the Faculty of the Harvard Business School early in the series of the Visiting Professors Case Method Programs. Reports of these discussions were given to participants in the summer Programs as a means of showing that the goal was not to answer all questions but a process of continuous approximation by which more useful questions could be discovered.

The sources of the questions raised by men with long careers in the case complex of activities merit brief review. Some differences in any discussion are, of course, semantic and might have been clarified by identifying the references. One faculty member might be talking about the situation in the field while another was talking about the situation in the classroom. Questions also arose from differences in emphasis on objectives. The questioner, for instance, might have in mind the need to develop a student's analytical skill while the speaker was stressing the need to develop a student's willingness to take action, or vice versa. The questioner might have in mind the need for students to consider the basic instruments of business when the speaker was stressing the need to identify purpose. Most would agree that all of these objectives are important for the administrator, but there is a great range of difference in what an individual faculty member would emphasize in his course or expect students to acquire in their total program. There are even sharper differences as to the sequence in which different objectives should be sought.

Furthermore, questions arise in the heat of controversy, because different possibilities for personal growth and development are hard to distinguish from pathological deviations. Differences in what is important to the individual along his own pathway to maturity are part of life and normal growth, for faculty as well as students. But pathology is possible when stresses are not kept in relationship. Professors with long experience in the case complex are aware of this and that is another reason for references to Appendix 4. Much heat came when a speaker pleading for analysis was interpreted as justifying the pathology of "indecision." Or when the discussion leader pressing to get students to make a decision was interpreted as favoring the counter pathology of impulsive action. Those who seemed to overstress the importance of skill were seen as condoning the pathology of "manipulation." Or those who appeared to be stressing the importance of situational thinking in dealing with ad hoc situations were characterized as "anti-knowledge" and condoning prejudice and arrogance. A school of business administration is concerned with an exceedingly complex process of growth from immaturity to maturity. It has responsibility to see that students grow in many dimensions. It is exceedingly important, therefore, to distinguish between essential but different elements in normal growth and the pathological distortions of growth in any one of these dimensions. These faculty discussions demonstrated how the isolation of one element even for attention and discussion departs enough from living interdependence to make Dean Donham fond of stressing the poet's realization that "our stability is but balance."

In the Program the fact that there were no students and that all the participants would leave within eight weeks provided a rare freedom to argue without need for reaching agreement on degree requirements or promotion criteria. The Program was expendable, unlike a continuing academic community in which members must find some way of working together while holding different opinions on vital issues.

Institutional Rewards and Punishments

In brief, the Program demonstrated that men from greatly differing backgrounds and institutions had the capacity to understand and to develop skill in the whole complex set of case-course development activities. Further, the experience of alumni of the Program demonstrated that skills could be carried on fruitfully in a broad cross-section of educational programs and institutions. The significant factor to consider, therefore, in the different rates of progress in study of administration by the use of cases is the institutional context.

The way the institution rewards performance is more complicated than the controversial dictum, "publish or perish." Case research, writing, and use needs to be incorporated into the institutional system of reward and punishment. The extent to which the case is or is not considered research; the extent to which the analytical skill developed in a course is recognized as an important objective; the extent to which understanding the administrative point of view is thought to be worth a professor's effort — are all institutional judgments of critical importance to a professor's career. Happily, an increasing number of schools have been finding ways to deal with these important conditions. The place of case research is only one element in the promotion policy and administrative structure of a school, and there have been many other issues of educational policy in these years calling for reappraisal of objectives and faculty organization. Some of these developments were discussed in Chapter VI and the case complex was seen as a means of helping faculties make progress toward new goals.

As the man in charge of a particular course, a professor enjoys great freedom. In his second role as a member of the academic community, he may exercise considerable influence and be correspondingly involved. The individual professor must live within the basic structure of his institution. It affects the extent to which his interest in administration may be given freedom for expression, even if he is not supported by concurrent action of colleagues or even by financial provision for travel, secretarial help, and time for getting into the field. The alumni of the Program have shown that it is possible for an individual professor to create and conduct his course to help his students develop for professional careers in business administration, even though the rest of the faculty are at the most permissive, if not actively hostile.

If a faculty has embarked upon development of a professional school or a professional program, there will inevitably be a high rate of interaction among faculty members regardless of teaching methods selected. The very minimum of determining objectives, and of selecting students, faculty, curriculum, courses, and methods involved in an integrated program, means that faculty members will be limited as well as supported by common decisions. The wear and tear on individual faculty members in developing and executing an integrated program for a group of students is a price which many faculties decline to pay. It also is easy to understand how faculty members who have lived through abortive attempts or even successful attempts at interdisciplinary programs react with indifference or even hostility to attempts of other faculty members to move them toward a collective goal.

Assuming, therefore, the probability that the individual professor will find himself in an academic community dominated by the unwritten code of "live and let live," it is important to note the points at which his adoption of a case method pattern may impinge upon or require some understanding from his colleagues and university administration. One of them appears in a predictable cycle of emotional involvement by students during a semester of the case method. Reactions typically run from initial excitement, into discouraging bewilderment, on to an enlarged sense of competence for handling administrative situations. In an integrated professional program where university administration and faculty are endeavoring to capitalize collectively upon the phasing of this emotional cycle to keep student interest high and creative, the individual instructor can experiment out at the limits of emotional involvement. In a hostile environment where the administration is afraid of student reaction, and colleagues are jealous of unusual student enthusiasm, it is not unnatural for the instructor to keep within safe limits of variation. How far he can go under any circumstance, of course, depends upon his individual skill and the position he has established in the academic community.

These facts of life might be sufficient to discourage innovation if it were not for the record of creative possibility demonstrated in a wide variety of institutions where individual professors have enlarged their own perspectives and clarified their professional objectives, while using the case method of studying business administration.

The basic responsibility of the administrator is to identify goals and policy directions within the administrative situation. His orientation is to the point of action. Similarly, when the professor becomes aware of choices that he must make in the study of specific administrative situations, he finds new stimulus. Much can be done within most academic institutions to get more of this stimulus from the study of administration by the case method. Fortunately, most schools allow a faculty member wide latitude in his selection of materials for a course. He is allowed freedom to experiment at least in his own course, even though he may not get financial support for extensive field research. The spread of MBA programs has helped to increase interest in materials for giving graduate students a better understanding of the administrative situation into which they are going. This interest is fostered by the corporations who are seeking MBA graduates and by their expectation that these men will have had some exposure to case situations. These influences also are leading to recognition that attention to administration as a process is important in addition to expertise in different departmental areas. The professor who has established his reputation as an expert in his area of specialization is rarely discouraged and often encouraged in developing an advanced course using cases which emphasize the management aspects of his subject.

Materials for experimentation are now readily available. The Intercollegiate Case Clearing House has received more than 7,000 cases which have been drawn from all parts of the country and designed for use at all student levels. The Intercollegiate Bibliography also lists over 220 books of cases and combinations of cases and text available from commercial publishers.

Finally, and of most importance, is the widespread willingness of American business firms to cooperate with professors from schools of business administration in case development. In the last 15 years there have been literally thousands of senior business managers who have participated in executive development programs in which cases have been used. The case researcher cannot rely upon finding cases ready-made, but there is an increasing probability of favorable reception by business executives to professors seeking case development.

Where professors in a wide range of institutions have actively gone out to study administrative situations, enthusiasm has increased. Where professors did not get involved actively in study of administrative situations, discouragement ensued. Even a small amount of time and effort in the field can serve as a catalyst to enrich the interchange between students and professor. When the atmosphere in the academic community encourages this effort to develop a living case method pattern, resources are found in ingenious ways. However, development is blocked when the academic community disparages the objective of professional competence in business administration. The need for financial support is inherent, although nominal, in relation to the leverage these funds give to better education for business administration. Other conditions which characterized the Program and its predecessors were outlined in the first chapter and have been adapted to circumstances in a widely varying range of institutions.

Maintenance of this multiple involvement which characterized the Program was sometimes difficult when men returned to their schools, because there was "no payoff" within their institution. The professor needs to be visible to a peer group whose evaluation confirms the wisdom of the university administration in his selection and promotion. Probably not enough emphasis has been put upon preparing journal articles from case situations by those who have become excited about the study of administration by cases. A researcher should not stop with the thrill of finding in a complex enterprise that his concepts help him and the active administrator discover the data which involve a critical issue for policy direction; nor even when students suddenly find that they too are gaining power to discover in the data of the real world a point where disciplined study makes a difference in their capacity to act. The case *is* where the action is. And the excitement of finding it in the field and of involving students in the classroom makes a person enthusiastic for more cases.

It may seem paradoxical, in a book about the case method, to stress the need for writing articles as well as cases. But this fact should be emphasized. In Chapter III on *The Written Case,* attention was called to the need to separate field data from ideas about them. The utility of collecting these ideas and commenting on them, as well as writing the case, was stressed. From these commentaries may come an article which restates the generalizations and hypotheses of a subject to help the researcher as well as the practicing administrator discover the emerging conditions out of which new directions and opportunities may be formulated. Such articles might well reflect the author's experience with the administrator — and the teaching of administration at the artistic, creative end of a scientific discipline. Concepts and methods which sharpen observation, interviewing, and interpretation may help make the relationship of administrator-professor mutually more helpful than assuming that the practitioner is only at the logical application end of the research process.

Detached Participation — The Coaching Function

While pointing out the merits of involvement and especially multiple involvement, the historical wisdom of detachment should not be forgotten. The Program coaches played an essential role in encouraging participants to get involved in multiple activities and to progress by interpreting and reconciling their reactions, thus demonstrating the power of the case method pattern. However, this role has to be understood in order to make the experience of the Program as an operating model fully transferable.

Appreciation for the work of these coaches should be repeated, since their contribution has never been adequately recognized. [See Appendix 2 listing the coaches for the 11 Programs.] Their work emphasizes the inherent task of interpreting different points of view characterizing the study of business administration — in the field and in the classroom.

Getting primary data of business administration means *personal involvement.* The coach helped to make such involvement more visible and to interpret its impact on getting data in the field and on learning in the classroom. The coach not only helped at the start of the Programs but repeatedly throughout the sessions helped each person to distinguish between what was seen and heard from theories of what should have been there. Both the administrator and the visiting professors had their own theories. The coach helped each of them respectively to clarify and express abstract concepts in order to discover the unique conditions in a specific enterprise, and how they might affect the direction of growth in the future. The coach also helped the visiting professor record the facts in a literary structure which the active administrator could approve, and which students would accept as real.

The coach also helped each professor think about his abstract subject in terms which could be recognized in the things, people, and events of life in the particular enterprise he was about to visit. The coach talked to the business administrator about the professor's visit and obtained some ideas about possible changes in the direction of the enterprise and the stresses implied in the choice of new directions. The logistical feat of introducing the professors and administrators tended to overshadow the fundamental job of interpreting different points of view.

This process of interpretation by the coach also led professors to reexamine course and curriculum objectives in order to fit specific cases. The coach helped to bridge the gap between the world of course abstractions and the living world of particular things, people, and events. Discussions of teaching strategy and application of feedback from classroom experiments were only the dramatic episodes in this continuing relationship.

This process of interpretation was not a simple, one-time conversation. It took time to discover the order existing in what people were doing in the enterprise, and how they felt about what they were doing and about those with whom they were doing it. It took time to discover points of contact between order in the life of a given enterprise and order in the abstract subject which the professor had isolated for mastery. It took time to clarify the meaning of class reactions to case data and leadership style.

* * * * *

The case method of studying business administration extends and elaborates administration as an abstraction by examining where it is involved in "life's interaction with life," where knowledge of the subject exposes new conditions in a living situation and, at this adaptive moment, provides the possibility for a creative step forward. An openness to actual situations and to the data of administration is needed, an alertness to decision points as well as rigor in the study of systems of knowledge under controlled conditions. It is not a question of substituting one study for another, but of finding means of communication. The essence of the case method is the discovery of the point in an enterprise where a decision has to be made as to the direction of its development as well as to the means to that end. Access to this point of decision for the administrator is through involvement in a whole system of routines. The issue created by these concrete facts in the life of this enterprise interacting with its environment may be illuminated by the specialized knowledge of the inquiring researcher. But the values and direction of policy are a venture into the unknown as well as an application of knowledge about abstract means. It is orientation to this adaptive moment which characterizes the administrative point of view. It is the need to act responsibly at the moment of discretion which requires skill in situational thinking, as well as logical inference and practice in monitoring impulse by informed reason.

The prevailing isolation of these processes for academic consideration assumes that the resulting knowledge can be put together somehow later on, but that is an assumption which student revolt now calls into question. The discovery and use of cases is one means of making progress on all these processes simultaneously: of bringing the administrator, the specialist, and the next generation together "where the action is."

The case method set of activities and relationships is relevant to this critical division in society because the case grows out of and permits comparison of the professional orientation of the administrator and that of the professor. The case is the mutual discovery of a body of data requiring discretionary action, the result of looking at a concrete situation from these different points of view. To discover such a case the professor has to get out to where the action is and get involved with the administrator and his struggle to adapt an ever-changing enterprise to an ever-changing environment.

The case method is relevant also because search for the case allows access to a man's own personal identity as distinct from his professional orientation. Although the recognition and separation of one's own feelings and beliefs from what he sees and hears sometimes is painful, the process is essential to growth and development. It begins with getting the data in the field. It continues in writing the case in a format which makes the researcher identify and exclude his own opinions while reporting only what he saw and heard for confirmation by the administrator and others in the actual situation. Similarly, the use of the case provides access to the values and identity of students in the classroom for contrast and comparison as the class uses the concepts of the course to find direction and appropriate action in the concrete set of facts.

The case is relevant because its discovery comes from efforts of the professor to see where and how his abstract knowledge helps to expose the fixed and variable conditions in the particular system of systems which is the concrete knowledge of the administrator. The case data offer opportunity for reconsidering policy direction and implementation and thus permit the student to discover for himself the relevance of abstract concepts from this subject for the world of concrete action.

The case is relevant because it arises from and allows for exercise of skill in living with complex situations. No one is completely an administrator. Administration is a process to be understood and experienced by all. No one is completely learned. Learning is a process to be understood and experienced by all. No one is completely mature. Maturity is a process to be understood and experienced by all.

The case is not the end, but it is one point in the process of evolution. It is imperative that we study such points because it is here in the world of things, people, and events that direction emerges from personal choice interacting with blind determinants of abstract systems.

Appendixes
Index

Appendix 1

Visiting Professors Case Method Program

List of Participants, 1955-1965

Name	*Year*	*University Represented*	*General Subject Interest*
Agnew, Robert J.	1956	Pittsburgh	Management
Anderson, Hershel M.	1963	North Texas State	Policy & Accounting
Arellano, Richard G.	1965	Florida State	Finance
Austin, Ellis T.	1962	Fresno State College	Finance
Baklanoff, Eric N.	1959	Louisiana State	Finance
Barksdale, H. C.	1957	New York (Commerce)	Marketing
Barrett, Gerald A.	1957	North Carolina	Policy & Ad. Prac.
Basil, Douglas C.	1956	Northwestern	Prod. & Personnel
Bayless, J. Leslie	1959	Louisiana Polytechnic	Mktg. & Mgmt.
Beagle, Chauncey M.	1963	Colorado	Acct. & Tax
Bedrosian, Hrach	1965	New York (Commerce)	Psychology
Beedle, Arthur	1960	British Columbia	Accounting
Bennett, William R.	1960	Alabama	Marketing
Berliner, William M.	1960	New York (Commerce)	Management
Berman, Wilfred	1957	Dalhousie	Accounting
Bierman, Harold, Jr.	1958	Cornell	Accounting Theory
Blair, Jacob J.	1958	Pittsburgh	Policy & Labor
Bloom, George F.	1955	Indiana	Finance
Bodiya, Naser G.	1964	Detroit	Marketing

Name	Year	University Represented	General Subject Interest
Bonwich, William	1965	Saint Louis	Marketing
Boyd, Virgil	1957	Northwestern	Accounting
Brabb, George J.	1965	Washington	Statistics
Brett, Frederic A.	1959	Alabama	Accounting
Bridges, Francis J.	1960	Georgia State College	Management
Brown, Leland	1958	Tulane	Communications
Brown, William N.	1963	Texas	Policy & Mktg.
Bryan, Stanley E.	1960	Michigan State	Administration
Burns, Thomas J.	1964	Ohio State	Accounting
Burris, Edward C.	1959	Oklahoma State	Economics
Buskirk, Richard H.	1956	Oklahoma	Marketing
Buteux, Raymond D.	1960	New York (Commerce)	Marketing
Butterworth, J. D.	1956	Florida	Retailing
Cahill, James M.	1962	Manhattan College	Economics
Call, Rex V.	1961	Arizona	Marketing
Campbell, Donald B.	1956	Instituto Mexicano de Administraction de Negocios	Administration
Cannon, Norman S.	1963	Utah State	Economics
Cantu-Quiroga, Abiel J.	1961	Instituto Technologica y de Estudios Superiores de Monterrey	Marketing
Carson, David	1955	Harvard Business School	Marketing
Chapman, James E.	1959	Georgia State College	Management
Chen, Gordon K. C.	1961	Rutgers State	Personnel
Christensen, Edward L.	1961	Brigham Young	Communications
Clark, Lincoln	1963	New York (Graduate)	Mktg. & Mgmt.
Coleman, Raymond W.	1955	West Virginia	Production
Cooper, Homer C.	1962	Pittsburgh	Admin. & Human Beh.
Cox, Alfred A.	1964	North Texas State	Marketing
Crites, Dennis M.	1958	Oklahoma	Marketing
Cronin, Justin C.	1957	Boston College	Ind. Mgmt.
Cundiff, Edward W.	1956	Syracuse	Marketing
Cutler, T. H.	1964	Denver	Bus., Govt. & Educ.
Darbyshire, Leslie	1959	San Jose State College	Policy & Investments
Davids, Lewis E.	1963	Missouri	Bank Administration

Name	*Year*	*University Represented*	*General Subject Interest*
Davier, Maurice	1955	Virginia	Production
Davis, James H.	1960	Ohio State	Marketing
Davis, N. F.	1965	Atlanta	Labor
Day, William H.	1958	Tulsa	Marketing
Dell'Aniello, Paul	1964	Montreal	Finance
Dellenbarger, L. E.	1963	Oklahoma State	Finance
DeSpelder, Bruce E.	1963	Wayne State	Administration
Dodge, Robert E.	1959	Oregon	Marketing
Dufton, Charles H.	1955	Northeastern	Marketing
Dulan, Harold A.	1955	Arkansas	Finance
Dunham, Clive F.	1955	Mississippi	Control
Duval, R. Errol	1962	Bishop's	Ad. Prac.
Ehlers, Carrol W.	1962	Georgia State College	Marketing
Emblen, Donald J.	1955	Montana State	Accounting
Facer, Elden J.	1962	Utah	Finance
Fagan, Paul H.	1961	New York (Graduate)	Ind. Mgmt.
Farwell, Loring C.	1955	Northwestern	Finance
Faulkner, David M.	1957	Syracuse	Prod. & Personnel
Finn, Robert H.	1964	Georgia	Business Policy
Finston, Howard V.	1963	New Mexico	Small Business
Gadon, Herman	1965	New Hampshire	Ind. Relations
Garcia, Fred Lawrence	1961	Fordham	Finance
Goldberg, Milton S.	1959	Michigan State	Management
Golis, Eugene F.	1964	Idaho	Policy & Admin.
Gourlay, Colin C.	1963	British Columbia	General Management
Gray, Harold E.	1964	MSU Center	Human Relations
Greaney, Walter T.	1961	Boston College	Accounting
Hadley, Clausin D.	1955	Santa Clara	Statistics
Hailstones, Thomas J.	1963	Xavier	Admin. & Economics
Hargrove, Merwin M.	1957	Tulsa	Management
Harrison, Frederick W.	1958	Babson Institute	Finance
Harrison, Ike H.	1958	Texas Christian	Policy

Name	Year	University Represented	General Subject Interest
Henning, Dale A.	1956	Washington	Policy & Personnel Rel.
Herrick, Kenneth W.	1962	Texas Christian	Insurance
Hersker, Barry J.	1964	Miami (Florida)	Marketing
Heysinger, Jack D.	1963	Wichita	Law
Hicks, William T.	1956	Mississippi	Econ. & Bus. Admin.
Hoben, William J.	1964	Dayton	Accounting
Hodgetts, James C.	1962	Toledo	Mgmt. & Bus. Policy
Holmes, Roderick L.	1964	Baylor	Acct. & Human Rel.
Huff, David Lynch	1961	California (Los Angeles)	Marketing
Hughes, George David	1964	Cornell	Marketing
Hungate, Robert P.	1962	San Diego State College	Finance
Hutton, Clifford E.	1965	Tulsa	Management
Jacobsen, Lyle E.	1960	Stanford	Accounting
Jensen, Hans	1962	Alaska	Economics
Joerg, Frederick C.	1959	Duke	Economics
Johnson, Lawrence A.	1963	Massachusetts	Marketing
Johnson, Robert L.	1962	South Dakota	Finance
Johnson, Thomas G.	1961	Oregon	Marketing
Johnson, Thurlo F.	1964	Massachusetts	Business Policy
Katzenmeyer, Robert	1961	Akron	Accounting
Kaye, Norman J.	1959	Marquette	Mgmt. & Economics
Keller, Thomas F.	1963	Duke	Accounting
Kelley, Eugene J.	1956	Clark	Retailing
Key, Henry	1959	Texas Christian	Accounting
Kibarian, Barkev	1958	Northeastern	Management
Kindig, Fred E.	1957	Pittsburgh	Ind. Mgmt.
Kirk, Roy Vernon	1965	Rensselaer Polytechnic	Ind. Relations
Klasson, Charles R.	1965	Texas	Policy
Kniffin, Fred W.	1958	Indiana	Sales Mgmt.
Knoke, William A.	1960	Iowa	Marketing
Kolasa, Blair J.	1965	Duquesne	Psychology
Kolde, Endel J.	1955	Washington	Marketing
Kramer, Walter	1960	Notre Dame	Marketing
Krey, John W.	1956	Boston	Ind. Mgmt.

Name	Year	University Represented	General Subject Interest
Kroeger, Arthur	1957	Stanford	Mktg. & Advertising
Krogh, Harold	1959	Kansas	Finance
Krzystofik, Anthony J.	1965	Massachusetts	Accounting
Lazarus, Harold	1964	New York (Graduate)	Management
LeBreton, Preston P.	1956	Detroit	Mgmt. & Bus. Policy
Lenberg, Robert A.	1963	Washington	Mktg. & Inter. Bus.
Little, Wallace I.	1958	Washington	Mgmt. Training & Trans.
Logan, James P.	1960	Columbia	Administration
Longenecker, Justin G.	1959	Baylor	Ind. Mgmt.
McAdams, Alan K.	1961	Cornell	Accounting
Maffei, Richard B.	1957	Mass. Inst. of Technology	Econ. & Operations Res.
Markin, Rom J.	1964	Washington State	Marketing
Martin, Bernard L.	1964	Xavier	Marketing
Martin, Jack	1957	Denver	Management
May, C. Kennedy	1962	Western Ontario	Marketing
Mayer, Morris L.	1962	Alabama	Marketing
Meek, Dudley H.	1955	Harvard Business School	Adminstration
Megginson, Leon C.	1958	Louisiana State	Pers. Mgmt. & Ind. Rel.
Meinhart, Wayne	1965	Oklahoma State	Management
Miller, Charles	1955	Marquette	Marketing
Miller, Gordon J.	1959	Utah	Accounting
Miller, Robert E.	1956	Southern California	Mktg. & Inter. Econ.
Mills, Robert H.	1962	Michigan	Accounting
Mitchell, Charles L.	1964	British Columbia	Accounting
Mitchell, Wiley	1955	Kansas	Acct. Theory
Moore, Carl L.	1959	Lehigh	Accounting
Morris, Malcolm L.	1965	Indiana	Marketing
Morrissey, Fred P.	1955	California (Berkeley)	Finance
Mullen, James H.	1958	Temple	Ind. Relations
Nelson, G. Kenneth	1959	Pennsylvania State	Accounting
Nelson, J. Russell	1965	Minnesota	Finance
Nightingale, Edmund A.	1958	Minnesota	Transportation

Name	*Year*	*University Represented*	*General Subject Interest*
Numer, Frank P.	1959	Duquesne	Management
Olm, Kenneth W.	1960	Texas	Production
Overman, Glenn D.	1961	Arizona State	Admin. Practices
Park, John C. S.	1964	Clarkson Col. of Technology	Finance & Mgrl. Econ.
Parrish, Lawrence L.	1956	Connecticut	Ind. Admin.
Paulis, Robert J.	1962	Rhode Island	Ind. Mgmt.
Penz, A. J.	1957	Alabama	Acct. & Control
Perkins, Bernard D.	1957	South Dakota	Pers. Mgmt. & Bus. Policy
Peterson, James M.	1964	South Dakota	Law & Business
Pond, Samuel A.	1962	Stanford	Bus. Org. & Policy
Poston, Charles F.	1963	Akron	Finance
Prater, George I.	1963	Dartmouth	Accounting
Prince, Thomas R.	1963	Northwestern	Accounting
Rahdert, Karl G.	1961	Bowling Green State	Management
Randall, Clyde N.	1957	Utah	Accounting
Raymond, Robert S.	1958	Ohio	Marketing
Reagan, Sydney C.	1962	Southern Methodist	Real Estate & Econ.
Reddoch, James W.	1957	Louisiana State	Marketing
Rettig, Jack L.	1965	Oregon State	Personnel
Ringstrom, Norman H.	1962	Oklahoma State	Marketing
Robb, James F.	1964	Portland State College	Mktg. & Inter. Bus.
Robb, Seward E.	1965	Oklahoma	Management
Roddenbery, Thaddeus H.	1964	Worcester Polytechnic Inst.	Mgmt. Engrg.
Rogers, Jack D.	1961	California (Berkeley)	Management
Sarjeant, J. A.	1959	Toronto	Production
Saunders, Charles B.	1957	Kansas	Marketing
Schley, Courtney F.	1955	North Dakota	Management
Schnitzer, Martin C.	1963	Virginia Polytechnic Inst.	Economics
Schwarzschild, Stuart	1963	Georgia State College	Finance & Ins.

Name	Year	University Represented	General Subject Interest
Scott, Karl M.	1961	Wake Forest College	Management
Scotton, Donald W.	1959	Illinois	Marketing
Searles, Clair K.	1955	Toledo	Production
Seelye, Frank D.	1965	Idaho State	Human Relations
Self, Stanley Allen	1961	North Texas State College	Management
Shapiro, Edward	1962	Detroit	Business Policy
Sheehan, James G.	1960	Cincinnati	Marketing
Singer, Frank A.	1961	Massachusetts	Accounting
Smith, Albert B.	1956	Pittsburgh	Retailing
Smith, Theodore H.	1957	Montana State	Mktg., Fin., Pol.
Soltis, Charles	1961	Kent State	Admin. Prac.
Soltow, Lee C.	1959	Ohio	Statistics
Sord, Burnard H.	1958	Texas	Management
Spencer, Charles H.	1957	Indiana	Accounting
Sprouse, Robert T.	1960	California (Berkeley)	Control
Stettler, Howard F.	1958	Kansas	Accounting
Stellmacher, Herbert Bob	1961	Hawaii	Marketing
Stevenson, Harold W.	1960	Minnesota	Finance
Swearingen, Eugene L.	1958	Oklahoma State	Prod. & Mgmt. Training
Taflinger, Ancel G.	1956	St. Col. of Washington	Personnel & Policy
Tate, Curtis E., Jr.	1958	Georgia	Ind. Mgmt. & Human Rel.
Taylor, Dale H.	1965	Brigham Young	Control
Taylor, James D.	1960	South Dakota	Marketing
Taylor, Weldon J.	1962	Brigham Young	Educ. Admin.
Thayer, Lee O.	1961	Wichita	Admin. Practices
Thomas, David A.	1957	Cornell	Accounting
Thomas, William E.	1960	Illinois	Accounting
Thompson, Ralph B.	1958	Florida	Mktg. & Oper. Research
Tjersland, Tore	1956	Washington (St. Louis)	Production Mgmt.
Traxler, Ralph N., Jr.	1963	Emory	Business Pol.
Trillich, B. L.	1962	Kansas	Business His.
Valentino, Salvatore	1960	Creighton	Finance
Van Eck, L. Frederick	1964	Missouri	Management

Name	*Year*	*University Represented*	*General Subject Interest*
Vasquez, Stephen W.	1960	Saint Louis	Admin. Prac.
Vaughn, Donald E.	1962	Louisiana State	Finance
Vieth, Arthur G.	1956	Kansas City	Finance
Vincent, Vern H.	1958	West Virginia	Acct. & Mgmt.
Wagle, John	1961	Oklahoma State	Marketing
Walters, S. George	1956	Lehigh	Marketing
Ward, Richard J.	1957	Fordham	Economics
Warrick, Walter H.	1965	Northwestern	Ind. Economics
Wasley, Robert S.	1955	Colorado	Acct. & Bus. Policy
Wettlaufer, John J.	1957	Western Ontario	Pers. & Ind. Rel.
Whitaker, Gilbert R., Jr.	1962	Northwestern	Bus. Economics
Williams, Arthur L.	1963	Pennsylvania State	Insurance
Wilson, Samuel M.	1962	Temple	Management
Winter, Lothar G.	1963	Hawaii	Marketing
Wood, L. Curtise	1960	Wichita	Management
Wortman, Max S., Jr.	1965	Iowa	Ind. Relations

Appendix 2

Visiting Professors Case Method Program

Faculty

	1955	'56	'57	'58	'69	'60	'61	'62	'63	'64	'65
Bennett, E. D.	—	—	C	—	—	—	—	—	—	—	—
Brown, M. P.	—	—	C	C	—	—	—	—	—	—	—
Culliton, J. W.[1]	C	—	—	—	—	—	—	—	—	—	—
Dearden, J.	—	—	—	—	—	—	—	—	—	C	—
Donham, P.	C	C	—	—	—	—	—	—	—	—	—
England, W. B.	—	—	—	—	C	C	—	—	—	—	—
Frese, W. F.	—	—	—	—	C	—	—	—	—	—	—
Fuller, S. H.	—	—	—	—	—	—	—	C	—	—	—
Guth, W. D.	—	—	—	—	—	—	—	—	—	—	C
Harrison, I. H.[2]	—	—	—	—	—	—	—	C	—	—	—
Hower, R. M.	—	—	—	—	—	—	—	—	—	—	C
Hunt, P.	C	—	—	—	C	—	—	—	—	—	—
Kennedy, M. T.	—	—	—	C	C	—	—	—	—	C	—
Kindall, A. F.	—	C	C	—	—	—	—	—	—	—	—
Lawrence, P. R.	—	—	—	C	—	—	—	—	—	—	—
Levitt, T.	—	—	—	—	—	—	C	—	C	—	—
McGarrah, R. E.	—	—	—	—	—	C	C	—	—	—	—
McNair, M. P.	—	R	—	—	—	—	—	—	—	—	—
Merry, R. W.	—	—	—	—	—	—	—	—	—	—	C
Nickerson, C. B.	—	—	—	C	—	C	—	—	C	—	—
Nolen, W.	—	—	—	—	—	C	—	—	—	—	—
Raymond, T. C.	—	—	—	—	—	—	D	—	—	C	C
Salmon, W. J.	—	—	—	—	—	—	—	—	C	C	—
Seiler, J. A.	—	—	—	—	—	—	C	—	—	—	—
Tilles, S.	—	—	—	—	—	—	—	C	—	—	—
Towl, A. R.	D	D	D	D	D	D	—	D	D	D	D
Tucker, F. L.	—	—	—	—	—	—	C	C	—	—	—
Vandell, R. F.	—	—	—	—	—	—	—	—	C	—	—
Wrapp, H. E.	—	C	C	—	—	—	—	—	—	—	—

[1] James W. Culliton was then Director of the Program in Business Administration at the University of Notre Dame.

[2] Ike H. Harrison was then Dean of the School of Business Administration at Texas Christian University.

Other coaches were from the Faculty of Harvard University, Graduate School of Business Administration.

Note: C=Coach; D=Director; R=Retailing Group.

Appendix 3

Visiting Professors Case Method Program

Outside Speakers

Dean G. Leland Bach
School of Industrial Management, Carnegie Institute of Technology
1955 — "Limits of the Case Method"
1958 — "A Friendly Critic Looks at the Case Method"

Sherwin Badger
New England Mutual Life Insurance Company
1964 — "Research Needs in Business Finance"

President John C. Baker
Ohio University
1957 — "Administration and International Relations"
1959 — "Some Thoughts on Affairs Abroad"
1960 — "The Role of Wallace B. Donham in the Development of the Case Method and Professional Education for Business Administration"

Professor Robert Bales, Director
Harvard University, Laboratory of Social Relations
1955 — "Observations"

Professor Donald F. Blankertz, Director
Graduate Division, Wharton School of Finance and Commerce, University of Pennsylvania
1959 — "Limitations of the Case Method"

Associate Dean F. Virgil Boyd
School of Business, Northwestern University
1965 — "A Case in Case-Course Development"

Professor Kingman Brewster
Harvard Law School
1959 — "Development of the Case Method in Problems of International Legal Significance"

Dr. C. Sidney Burwell
Harvard Medical School
1958 – "The Case Method of Instruction in Medical Education"

Professor Nathaniel Cantor
University of Buffalo
1955 — "What is Learning?"
1956 — "The Teaching-Learning Process"
1957 — "Teaching-Learning Process"

Dr. Thomas Carroll, Vice President
Program in Economic Development and Administration, The Ford Foundation
1960 — "General Discussion of Education for Business Administration"

____________________, President
George Washington University
1962 — Background of the Program and education for business responsibility

Dr. Neil W. Chamberlain, Director
Program in Economic Development and Administration, The Ford Foundation
1959 — "A General Theory of Economic Process"

Dr. James W. Culliton, Director
Program for Administrators, Notre Dame University
1955 — "Undergraduate Programs for Administrators"

____________________, Dean
College of Commerce, Notre Dame University
1957 — "Toward a Theory of Business Administration"
1958 — "The Case of the Conceptual Scheme"
1959 – "The Teaching-Learning Use of the Case Method"
1962 — Experiments in learning to write cases and an experiment in developing a program using cases with undergraduates
1965 — "Substance Versus the Form of the Case Method"

Professor Maurice Davier (SCSP 1955)
Graduate School of Business Administration, University of Virginia
1960 — General seminar discussion

Associate Dean William Dill
Carnegie Institute of Technology
1965 — "The Carnegie Tech Concept of Education for Business Administration"

Dean Richard Donham
School of Business, Northwestern University
1958 — "What's Going on Here?"
1959 — "What's Going on Here?"
1963 — "The Drexel Electronics Company Case and the Case Method in General"

Professor Charles Dufton (SCSP 1955), Head
Marketing Department, Northeastern University; President, Boston Chapter American Marketing Association
1957 — Organizing for case collection; the use of cases in developing undergraduate courses

Professor Albert H. Dunn, III
University of Delaware
1957 — "Experiment in Case Research in a Liberal Arts College"

Dr. Wallace Feurzeig
Bolt, Beranek and Newman, Inc.
1964 — Computerized case discussion

Dean Paul Garner
School of Commerce and Business Administration, University of Alabama; President, American Association of Collegiate Schools of Business
1964 – The task of the professor in a school of business administration facing the problems and opportunities now developing in the United States and around the world

Professor Billy Goetz
Massachusetts Institute of Technology
1956 — "Use of Models in Teaching Business Administration"
1957 — "Production and the MIT Program in Industrial Management"

Dr. William P. Gormbley, Jr., Associate Director
Program in Economic Development and Administration, The Ford Foundation
1963 — "The Issues Facing Education for Business Administration"
1964 — "The Tasks of Professors and Schools of Business Administration"

Dr. Edward T. Hall
Washington School of Psychiatry
1962 — "The Contribution of Anthropology to the Analysis of Cases in Different Cultures"
1963 — "Culture, Administration, and the Case Method"
——————, Professor of Anthropology
Illinois Institute of Technology
1965 — "Organizational Culture and the Case Method"

Dr. Oscar Harkavy
The Ford Foundation
1957 — "Implications of the Summer Case Seminar Program for Improving the Teaching of Business Administration"

Dean Ike H. Harrison
School of Business, Texas Christian University
1963 — "The Implications of Behavioral Science for the Case Method"

Dean Karl Hill and Associate Dean John Hennessey
Amos Tuck School of Business Administration, Dartmouth College
1964 – "The Dartmouth Point of View; The Relationship of the Professor to his Institution"
1965 — "Institutional Influences on Use of the Case Method"

Professor Stanley Jacks
Massachusetts Institute of Technology
1960 — Recording of the Mount Hope Finishing Company Labor Relations Dispute of 1952-1954

Dean Howard Johnson
Alfred P. Sloan School of Industrial Management, Massachusetts Institute of Technology
1962 — "The MIT Program of Education"

Dr. Horace Kallen
New School of Social Research
1955 — "A Philosopher's View of Learning"

Professor Abraham Kaplan
College of Literature, Science and the Arts, The University of Michigan
1964 — "The Conduct of Inquiry"

Dean Francis Keppel
Harvard Graduate School of Education
1957 — "The Teaching-Learning Process at the Secondary Level"
1959 — "Current Issues in American Education"

Dr. Edwin Land, President
Polaroid Corporation
1965 — "Utilizing the Learning Process in Industrial Research and Development"

Acting Dean James Lorie
School of Business, University of Chicago
1959 — "The Philosophy of Business Education with Particular Reference to Financial Management and How to Teach It"

Dr. David C. McClelland, Chairman
Department of Social Relations, Harvard University
1962 — "Achievement Motivation and Education for Business Administration"
1964 — "Achievement Motivation"

Professor Warren McCulloch
Massachusetts Institute of Technology
1955 — "Memory"

Professor Douglas McGregor
Massachusetts Institute of Technology
1958 — "Organization for the Management of Human Resources"

Dr. Kiyo Morimoto, Associate Director, and Dr. William G. Perry, Director
Harvard Bureau of Study Counsel
1965 — Taped tutorial discussion — example of teaching style

Dean Russell D. Niles
School of Law, New York University
1958 — "The Use of the Case Method in Legal Education: Cases from the Courts and from Raw Facts"

C. Northcote Parkinson
1958 — "The Evolution of Political Thought"

Dean R. F. Patterson
School of Business Administration, University of South Dakota; President, American Association of Collegiate Schools of Business
1963 — "The Problems of Educating for Business Administration"

Professor and Mrs. Paul Pigors
Massachusetts Institute of Technology
1957 — "The Incident Method"
1962 — "The Incident Method"
1963 — "The Use of the Incident Method"

Irving Rabb, Vice President and General Manager
Stop and Shop, Inc.
1962 — "The Industrial Enterprise as an Institution of Continuous Learning"

Dean Philip Ragan
College of Business Administration, Boston University
1959 — "Plans for Business Education at Boston University"
1962 — "Problems in Administration of a Metropolitan School of Business Administration"

Dr. Marshall Robinson, Director
Program in Economic Development and Administration, The Ford Foundation
1965 — "Business Administration: The Ford Foundation's Present Involvement and Prospects for the Future"

Professor Herbert A. Simon
Graduate School of Industrial Management, Carnegie Institute of Technology
1960 — "Basic Mathematics"

Professor Burnard Sord (SCSP 1958)
University of Texas
1959 — "The Southern Case Writers Group"

Dr. J. Richard Suchman
University of Illinois
1964 — "A Demonstration of Inquiry Training"
________________, Director
Curriculum and Demonstration Program, Office of Education
Department of Health, Education and Welfare
1965 — "Inquiry Training: The Autonomous Discovery of Relationships Between Data, Theory and Action"

Dr. Renato Tagiuri, Research Associate
Laboratory of Social Relations, Harvard University
1956 — "Small Groups and the Learning Process"

Mr. Antonio Valero, Director
Instituto de Estudios Superiores de la Empresa
1960 — "Recent Developments in Spanish Industry"

Dr. Bert L. Vallee
Biophysics Research Laboratory, Harvard Medical School
1959 — "The Relationship Between Theoretical Models and the Case Method"
1962 — "Medical Science and the Case Method: Analogies in Business problems of Integrating Mathematical Models, Behavioral Sciences and the Case Method"

Dean W. Allen Wallis
School of Business, University of Chicago
1958 — "A New Program for Education in Business Administration"
1962 — "Alternative Approaches to Education for Business Administration"

Dr. Lewis B. Ward, Director of Research
Educational Testing Service
1955 — "Executive Study: Selection"
1956 — "Executive Selection and Development"

Dean Samuel Westerfield
School of Business, Atlanta University
1960 — "Observations of the Case Method at the Harvard Business School and Plans for Studying the Needs and Opportunities for Management Education"

Professor Edward Wrapp
Graduate School of Business, University of Chicago
1964 — "The University of Chicago Approach to the Study of Business Administration in Its Executive Development and Its Degree Programs"

Associate Dean John Wynne
Alfred P. Sloan School of Industrial Management, Massachusetts Institute of Technology
1963 — "The Business Programs of M.I.T."

Appendix 4

Visiting Professors Case Method Program

Harvard Business School Faculty Case Method Seminars, 1959-1960

Summary

The participants in every Program expressed surprise that there were so many differences among members of the Harvard Business School Faculty in their views of the case method. By the end of each summer, they saw that there were indeed as many "case methods" or patterns as there were professors. The following summary of a special series of seminars held by the Faculty of the Harvard Business School during 1959-1960 allowed further consideration of these differences. The plan for the Faculty Seminars was to have four professors present their ideas on the case method during the fall of 1959, to be followed by a series of open discussions during the winter of 1960. The topics chosen were:

"Conditions which Make Use of a Case an Effective Tool"
Professor Sterling Livingston

"Integrating Knowledge and Skill in a Case Course"
Professor Paul Lawrence

"The Case Method as a Normative Theory of Administration"
Professor Paul Cook

"Purpose and Problems in Using Integrated Case Material"
Professor John Glover

In analogy with Harvard's traditional case footnote, these particular professors were asked to present their views for the purpose of promoting discussion and not to illustrate either effective or ineffective practice of the case method. Neither can they be held accountable for the inevitable distortion of their views from the condensed notes, which cannot do justice to their thought-provoking statements and the discussions which they aroused.

Conditions which Make Use of a Case an Effective Tool

For the first of the meetings led by Professor Livingston, participants were asked in advance to analyze the Vigard Company case. He explained that he had prepared a taped recording of a training program in which discussion leaders were brought together to participate in a demonstration of the case method. Professor Livingston wanted the Harvard Faculty seminar to consider three questions in listening to the taped case discussion:

(1) Is this a typical case discussion?
(2) What is the instructor trying to accomplish in this case discussion?
(3) Has it been accomplished?

After playing the introductory talk about the case method, Professor Livingston stopped the tape to explain, "In the early Spring of 1956, Mr. Jesse Krause, Purchasing Agent of the Vigard Manufacturing Company, was faced with the problem of selecting a source to prepare a technical manual for a winder that the company was planning to sell nationally."

Professor Livingston then declared, "I find it useful to have men commit themselves at the outset of a case discussion." He thereupon went to the board and listed the four alternative sources from which bids had been received. There was a good deal of noisy protestation, but eventually he was able to tally votes on the board.

Obviously, the Harvard professors did not want to be committed. In the ensuing outburst of question, comment and reaction, Professor Livingston acknowledged that he wanted his students to be thinking about the selection of sources. He was trying to get across his concept of source selection. This didn't mean, in his opinion, that the instructor should seek to lead the class to any one choice. Although Professor Livingston was eager to get attention back to the taped discussion of the case, his repeated suggestions about going ahead with the tape and listening to the techniques that were being used by the instructor were interrupted by additional questions from the floor for about 45 minutes. Professor Livingston's explanation that forcing students to a conclusion, in his mind, meant taking a vote of where they had come out, stirred further questions. For instance, what do you want the students to decide on? Are you concerned about the relation of the course of action to conditions? Another member of the Faculty wanted a tally of how many men didn't want to vote.

Professor Livingston said at this point, "It is probably clear to you that I want the class to discuss what I think is important."

Many in the seminar were still restive from being "pressured to vote." They tried to get attention to decision making as a social process as well as an intellectual process. Or they wanted to examine the evidence actually in the case, feeling either that the case lacked the facts or that the leader of the taped discussion had not really brought out the basic considerations in a source selection issue.

Professor Livingston's main propositions which seemed to provoke so much reaction are reproduced here:

(1) It is important for an instructor to make sure that a student reaches a decision.

(2) It is important for students to learn the consequences of their decisions so they will know when they have made mistakes; i.e., students must not make wrong decisions without knowing they are wrong.

(3) It is important for an instructor NOT to lead a student to any particular decision. If the instructor leads the students to a predetermined decision, he manipulates. He must allow students to make mistakes.

(4) However, when an instructor merely questions, restates and summarizes, the discussion is not entirely satisfactory; because students may be only voicing and rationalizing their prejudices instead of developing a soundly reasoned decision from analysis.

(5) Therefore, we must have *B* cases to confront students with the consequences of their decisions.

In reviewing this seminar discussion five years later, Professor Livingston reiterated his conviction and expressed concern that his point of view put responsibility upon the instructor himself to follow up in the field what happened as a sequel to the original case situation. By comparing his own decisions with what had happened in the actual situation, he felt that he himself had learned a great deal in the years following the Faculty seminars.

Integrating Knowledge and Skill in a Case Course

Much discussion in the first two meetings had been around what had been accomplished in the discussion of a particular case and what could be accomplished in a course as a whole. The next session, under the leadership of Professor Paul Lawrence, began by his talking about the problems of integrating knowledge and skill around the outline for the Administrative Practices course.

(1) Our course objective is to increase professional capacity which involves both knowledge and skill.

(2) We work toward this objective through the discussion of cases, development of new kinds of cases, and new series of cases.
(3) We are experimenting with how to feed in knowledge through research readings, notes on analytical tools, and lectures.
(4) We are experimenting with development of personal skills through simulation exercises, and occasional discussion of what is taking place in class while cases are being discussed.
(5) The instructor has many different roles, such as: being directive in drill situations, calling for a commitment of the student to action in action situations; and providing research data on foreseeable consequences to student proposals.
(6) Our experimentation has demonstrated potentials as well as problems and by getting reactions to these, we are trying to continue development of the course.

In his presentation, Professor Lawrence acknowledged that he shared the great perplexity of Faculty members in the previous discussion who had asked, "Where does knowledge get into the case method?" He and his group had seen that different approaches required different treatment on the part of the instructor. Among the assets of the case method Professor Lawrence noted that use of cases provided opportunity to hold together all facets of the problem of communicating knowledge and the development of skills. The liabilities, however, were illustrated by the pressures upon the Faculty when shifting to the role of drill sergeant to get across certain fundamental knowledge. Professor Lawrence noted that it was difficult to make this shift without confusing the students, who had earlier been encouraged to express their latent attitudes toward authority or toward the instructor himself by the instructor's very permissive role in class.

He pointed out the difficulty of getting students to deal with conceptual schemes. They would tend to ignore the ideas when confronted with a particular case or would veer the other way and apply the conceptual scheme to a particular case whether it was relevant or not. This discussion of conceptual schemes for case analysis prompted Professor Livingston to get to his feet and repeat his emphasis upon getting students to make a decision. He felt that the instructor's emphasis upon analysis "brings indecision." He wanted students to have arrived at a decision before coming to class and then to reappraise and reconsider their personal decisions during the class discussion. He felt that, as long as they had made a decision, his responsibility as the instructor was only to see that students became aware of the consequences of their decisions by facing a second case showing the sequel to the first situation.

Professor Lawrence used the intense discussion of getting a decision and the response to errors to illustrate his point that the instructor must play different roles at different times. Much discussion turned on the meaning of *decision* to different instructors. One thought of himself as the manager of a company when he was leading the discussion of a case. Several professors felt their role was quite different. Obviously the meaning of *decision* had divergent implications for the way they would conduct a class and the way they would organize a course. Another dimension of the discussion had to do with *sequence* in the introduction of knowledge and cases. Some members of the Faculty felt that knowledge should be stressed first and cases introduced thereafter. Other members of the Faculty felt just the opposite, while Professor Lawrence expressed his personal preference for mixing.

The Case Method as Normative Theory of Administration

Professor Cook met with the Faculty Case Method Seminar on November 21, 1959 and went directly into an exposition of the case method as the substance of administration rather than as pedagogy. His ideas in outline were:

I. The case method is not merely pedagogy; to a great extent it is substance. The substance appears in the creation in the student of the administrative point of view.

II. APV [the administrative point of view] is a normative theory of decision-making; despite vagueness and generality, it incorporates our idea of how decisions ought to be made:

- A. With an action orientation
- B. Taking account of the full range of relevant considerations--human, economic, environmental, functional, etc.
- C. Reflecting the "state of the arts" of analysis

Thus students should learn to deal effectively with administrative "reality," in the sense of total problems, using properly and when applicable the most effective analytical tools.

III. Cases and case discussions do not, however, simulate reality; they overemphasize analysis and foreshorten time, so that issues unresolved by business over a period of years tend to be treated as if they were decided in 80 minutes.

IV. Admitting the "state of the arts" of analysis as something to be learned has two major effects which raise major problems:

A. Teaching becomes directive; the instructor has superior knowledge which he seeks to impart to students. Query: Is the case method then only a bag of tricks, method without substance?

B. Analysis tends to be "partial, i.e., to break out of the context of a total administrative problem one segment of that problem which can be analyzed with specialized tools. Query: Since such "partial" analyses do not reflect total administrative problems, is a tendency to teach them a centrifugal force operating on the overall objective of giving the APV to students?

V. The "partial" analyses do not conform to the functional fields of business; they are generally academic disciplines which abstract from total problems some elements which are amenable to certain kinds of analysis.

VI. Such teaching is practiced here more widely than either critics or friends have stated. It appears to greater or lesser extent in all first-year courses, except perhaps WAC [Written Analysis of Cases]. It is, however, resisted and compensated for, and students generally are not deceived into believing that what is being learned is "the" way to make decisions. This is an inevitable result of having young men teaching, who are anxious to be at the forefront of their fields. The "wise man" type of mastery of a field is generally not available to young men.

VII. Devices personally used to keep administrative perspective are as follows:

A. There is always a question asked to which I don't know the answer; analysis is subsidiary to the approach to that question.

B. The initiative to learn is kept on the student.

C. Subcases which broaden the issues are

introduced and briefly discussed in class.

D. A decision is required, to provide focus for discussion.

VIII. A heavy load is placed, however, on second-year courses and Business Policy, which must not only integrate functional fields--which is relatively easy--but must integrate the various "parts of decisions" that students have learned to analyze separately.

IX. Another major concern is obsolescence and deterioration through lack of use of the content taught and the knowledge imparted: What is the best way to teach a "theory" or a "conceptual scheme" or a "set of facts" when we recognize these limits? If we are not, don't want to, and cannot be, training professional business economists, behavioral scientists, etc,, what is appropriate to teach from these fields? Do we teach merely that they exist, or what they have found? A tentative (and not wholly satisfactory) answer is that the appropriate specialties of our students are those they acquire after they leave here, and that the role of our teaching "partial" analyses ought to be to equip students better to learn such specialties and apply them using all their administrative abilities.

Some of the senior members of the Faculty commented during the discussion and later that the typical case method instructor tended to "think like a generalist while acquiring the data of a specialist." Examples given were Professor Neil H. Borden, who thinks like a generalist about the problems of business as a whole but knows the data of advertising. Or Professor Franklin Folts, who thinks like a generalist about the problems of business as a whole but knows the data of the factory. Or Professor Malcolm P. McNair who thinks like a generalist about the problems of business management but has a wealth of background, experience, and data on retail distribution.

Professor Cook precipitated further discussion by posing a distinction between the typical business school subject which had emerged in some rough correspondence with specialization in the management of an enterprise, such as advertising, production, or retailing, and the academic disciplines which were emerging from research abstractions about the behavior of man or of mathematical systems. In effect, the question was posed whether, in future generations of the Faculty, men could still think like generalists about the management of an enterprise while acquiring the data of economic analysis, of organizational behavior, of computer technology, and quantitative analysis.

Professor Cook's image of the case method as a normative theory of administration raised additional questions. He pressed his point by calling the Faculty's attention to the fact that there is something which the instructor knows that the students do not know at the beginning of the term. If the student hasn't learned this by the end of the term, the student does not "pass." In the case method this something is the ability to look at the situation as a whole and to decide on an appropriate course of action in this particular case. Expanding the term "case method" to "the case method of instruction" disturbed Professor Cook because this implies that the case method is just a bag of tricks to get students to acquire another body of information. He stressed that in analogy with the scientific method, the case method is the method of dealing with administrative situations. To get across the administrative point of view, this normative way of dealing with administrative situations is a central responsibility of the Faculty.

Purposes and Problems in Using Integrated Case Material

Professor John D. Glover at the time of his presentation to this Faculty Seminar on December 12, 1959, was Chairman of the Faculty in the Advanced Management Program, a 13-week residential program in which there were approximately 150 men from senior management positions, averaging about 45 years of age. The Faculty assigned to the program typically had this as their only responsibility for the term.

In the last seven regular class days of the program students (executives) were asked to concentrate upon Operation Phoenix. The class was divided into four sections and each section chose a "management consulting team" of five members who were charged with preparing recommendations to the company management. On four of the class days, all of the class met together with the entire teaching group sitting in the front of the room to discuss all aspects of the company situation from the top management point of view. Each of the four days covered a segment of time in the company's experience. Material about each new situation was given out at the end of the day, so that the group as a whole was dealing with an evolving situation.

Later in the sequence, competitors of the company appeared before the class to present their latest product lines. Each section met with its team to discuss the current situation, and to develop ideas and recommendations.

Then, each section management consulting team presented its report before the entire session, the Faculty, and representatives of the company management. Awards were announced by the management and Faculty to the team judged to have given the most outstanding report.

The underlying philosophy which led to this Operation Phoenix appears to have come from some such set of propositions as summarized below:

(1) Getting all the Faculty and all students involved actively in working out a decision in a project so big as to overshadow any special knowledge or skill has positive values.

(2) It is important to "tinker around" with varieties of teaching materials.

(3) It is disturbing that students (and colleagues) considered as good, management in a case which seemed to be "bristling with symptoms of degeneration.'

(4) The Faculty needs to accept as "homework" the follow-up of what happens in companies used as cases.

(5) Our difficulty is a failure in diagnosis, the reading of "the handwriting on the wall." The facts were there, but were not read.

(6) We need diagnosis, the relation of parts to the whole, rather than analysis, the mechanical breakdown of a subject.

(7) For the purpose of approaching reality, we need to develop cases providing facts evolving and allowing integration along three dimensions:

  (a) Integration over time

  (b) Integration over functions (horizontal)

  (c) Integration over different levels of organization (vertical)

(8) Too often our cases abstract out only one or a few of these essential dimensions, leading us and our students to a false sense of reality.

(9) Cases like the Phoenix series moving closer to the territory, to reality, actually have economies for the Faculty and students.

  (a) Time taken to master the basic company facts can be utilized in facing a variety of issues and their interrelationship.

  (b) Time to establish basic company contacts can save time in getting further cases.

(10) Cases like the Phoenix series raise problems which we as a Faculty ought to face as we try to see facts as an organically related system.

Professor Glover pointed out that the purpose of integration in Operation Phoenix was manifest by having all of the instructors in front of the class at once and subsequently in having the President and some of his management associates there also. This same integrating purpose had been manifest earlier in the preparation of case material when several different members of the Faculty undertook responsibility to see that data were included in the case which they felt were necessary to diagnose the situation in ways which were important to their subjects. The cases were thought of as steps toward bringing the real territory into the classrooms. The teaching group "had vowed we were going to be generalists." "Our AMP Faculty purpose was, regardless of our specialized interest, to see this as an organically related system."

Several professors were skeptical whether the effort to bring a massive collection of facts from the company into discussion by the entire program was any more effective than a simulation game. After all, the massive case was still "only an approximation." Everyone recognized that the presence of the management of the company and the competitive aspects of the end of the program gave for all of the Faculty and management participants a sense of importance.

Professor Glover and other members of the Advanced Management Program teaching group stressed their concern with diagnosis of the totality rather than a mechanical analysis of isolated parts. Nevertheless, questions kept coming up in the Faculty discussion whether the participants had found conceptual bases around which to integrate their comprehension of the management process. Or whether the participants had taken from the School only a sense of having worked closely with men of different interests around a great variety of materials which had been illuminated by Faculty people with different backgrounds under the stimulus of the actual management concerned for direction under stress.

Reference to further questions arising from this series of Harvard Faculty Case Method Seminars is made in Chapter VI.

Index